TP
440
.F78

Fruit
processi
edited b
Arthey
R. Ashu

Fruit Processing

Fruit Processing

Edited by

D. ARTHEY
Deputy Director
Campden Food and Drink Research Association
Chipping Campden
Gloucester
UK

and

P.R. ASHURST
Dr P.R. Ashurst and Associates
Kingstone
Hereford
UK

BLACKIE ACADEMIC & PROFESSIONAL
An Imprint of Chapman & Hall
London · Glasgow · Weinheim · New York · Tokyo · Melbourne · Madras

Published by
Blackie Academic & Professional, an imprint of Chapman & Hall,
Wester Cleddens Road, Bishopbriggs, Glasgow G64 2NZ

Chapman & Hall, 2–6 Boundary Row, London SE1 8HN, UK

Blackie Academic & Professional, Wester Cleddens Road, Bishopbriggs, Glasgow G64 2NZ, UK

Chapman & Hall GmbH, Pappelallee 3, 69469 Weinheim, Germany

Chapman & Hall USA, Fourth Floor, 115 Fifth Avenue, New York NY 10003, USA

Chapman & Hall Japan, ITP-Japan, Kyowa Building, 3F, 2-2-1 Hirakawacho, Chiyoda-ku, Tokyo 102, Japan

DA Book (Aust.) Pty Ltd, 648 Whitehorse Road, Mitcham 3132, Victoria, Australia

Chapman & Hall India, R. Seshadri, 32 Second Main Road, CIT East, Madras 600 035, India

First edition 1996

© 1996 Chapman & Hall

Typeset in 10/12pt Times by Cambrian Typesetters, Frimley, Surrey

Printed in Great Britain by The University Press, Cambridge

ISBN 0 7514 0039 4

Apart from any fair dealing for the purposes of research or private study, or criticism or review, as permitted under the UK Copyright Designs and Patents Act, 1988, this publication may not be reproduced, stored, or transmitted, in any form or by any means, without the prior permission in writing of the publishers, or in the case of reprographic reproduction only in accordance with the terms of the licences issued by the Copyright Licensing Agency in the UK, or in accordance with the terms of licences issued by the appropriate Reproduction Rights Organization outside the UK. Enquiries concerning reproduction outside the terms stated here should be sent to the publishers at the Glasgow address printed on this page.

The publisher makes no representation, express or implied, with regard to the accuracy of the information contained in this book and cannot accept any legal responsibility or liability for any errors or omissions that may be made.

A catalogue record for this book is available from the British Library

Library of Congress Catalog Card Number: 95–80923

∞ Printed on permanent acid-free text paper, manufactured in accordance with ANSI/NISO Z39.48-1992 (Permanence of Paper).

Preface

Fruit and fruit products, in all their many varieties and variations, are major world commodities and part of the economic life blood of many countries, particularly in the developing world. The perception of the healthy nature of fruit is a major reason for its increased consumption in the developed world, and many consumers today find a wider selection of fruit varieties, available at all times of the year, than ever before.

This volume, however, is not so much concerned with fresh fruit as those principal areas of processing to which it may be subjected. Fruit processing arose as a means of utilising a short-lived product and preserving its essential nutritional qualities as far as possible. A chapter on the nutritional aspects of fruit is included in this work to reflect the importance of this topic to most consumers.

After a general introduction, the chapter on fruit storage is the only contribution which deals with a process from which fruit emerges in essentially the same physical condition. Beyond that the book sets out to cover most of the major areas in which fruit may be processed into forms which bear varying semblances to the original raw material.

A chapter on juices deals with the biggest product area associated with fruit; there is hardly a country in the world that does not have a juice manufacturing industry. Many fruit juice producers go on to a second stage and ferment the juice to produce wine, cider and similar products. A chapter on wine production from grapes has been deliberately omitted because it is so large a subject and is already well documented. Despite that omission there is an important contribution on cider making which also covers the production of speciality fruit wines.

The subject of thermal processing of fruit, by both heating and freezing, is still important as a method of preserving or preparing fruit for various uses and is included, forming a comprehensive review of the preparation of preserves, flavours and dried fruit.

All fruit processing leads to the generation of significant quantities of unwanted material and an important industry has grown up around developing valuable products from such waste. The chapter on this topic covers some important materials including pectins and various citrus extracts. Regardless of the efficiency of a fruit process, there will always be an environmental impact and the final chapter deals with the subjects of effluent treatment and water, which is vital to so many fruit processors.

This book has had an unusually long period of gestation resulting in an authoritative contribution to a large world wide industry where companion volumes are difficult to find. The editors take full responsibility for the book, and any omissions and are particularly grateful to the authors who have contributed their work. We are confident the book will be of wide interest to technologists and those with commercial involvement throughout the world of fruit processing.

D.A.
P.R.A.

Contributors

B. Beattie	New South Wales Agriculture, Horticultural Research and Advisory Station, PO Box 581, Gosford, New South Wales 2250, Australia
R.W. Broomfield	78 Court Road, Malvern, Hereford and Worcester WR14 3EG, UK
G. Burrows	Hillsdown Holdings plc, Bridge Road, Long Sutton, Spalding, Lincolnshire PE12 9EQ, UK
A.L. Cohn	Ruth Cohn Rulek Ltd, Food Services, 33 Hanarkissim Avenue, Ramat-Gan 57597, Israel
R. Cohn	Ruth Cohn Rulek Ltd, Food Services, 33 Hanarkissim Avenue, Ramat-Gan 57597, Israel
P.C. Fourie	Stellenbosch Institute for Fruit Technology, Infruitec, Private Bag X5013, Stellenbosch 7599, South Africa
B. Jarvis	HP Bulmer Ltd, The Cider Mills, Plough Lane, Hereford and Worcester HR4 0LE, UK
P. Rutledge	CSIRO Food Research Laboratory, PO Box 52, North Ryde, New South Wales 2113, Australia
R.B. Taylor	Borthwicks Flavours Ltd, Denington Estate, Wellingborough, Northamptonshire NN8 2QJ, UK
N. Wade	New South Wales Agriculture, Horticultural Research and Advisory Station, PO Box 581, Gosford, New South Wales 2250, Australia
M.J.V. Wayman	First Effluent Ltd, 42a High Street, Sutton Coldfield, West Midlands B72 1VJ, UK

Contents

1. Introduction to fruit processing — 1
R.B. TAYLOR

1.1 Processing on a global scale — 1
1.2 Factors influencing processing — 2
1.3 Fruit types for processing — 3
   1.3.1 Pome fruits — 3
   1.3.2 Citrus fruits — 5
   1.3.3 Stone fruits — 5
   1.3.4 Soft fruits — 6
1.4 Controlling factors in the ripening of fruit — 7
   1.4.1 Respiration climacteric — 7
   1.4.2 Ethylene production — 8
1.5 Biosynthesis of flavours — 10
   1.5.1 Analytical data — 10
   1.5.2 Taste and aroma — 10
   1.5.3 Flavour formation — 11
   1.5.4 Physiological and biochemical aspects — 13
1.6 Factors influencing fruit quality and crop yield — 15
   1.6.1 Fruit variety — 15
   1.6.2 External factors affecting fruit quality — 16
1.7 Flavour characteristics — 16
1.8 The global market: threats and opportunities — 17
1.9 Fruit processing — 18

2. Fruit and human nutrition — 20
P.C. FOURIE

2.1 Introduction — 20
2.2 Composition of fruits — 20
2.3 The importance of fruit in the human diet — 32
2.4 Changes in nutritive value during processing — 35

3. Storage, ripening and handling of fruit — 40
B. BEATTIE and N. WADE

3.1 Maturing and ripening — 40
   3.1.1 Climacteric behaviour — 41
   3.1.2 Ethylene — 42
   3.1.3 Maturity standards — 44
3.2 Temperature and respiration — 45
   3.2.1 Field heat removal by pre-cooling — 47
   3.2.2 Cool storage of fruit — 51
   3.2.3 General requirements of a coolroom — 53
   3.2.4 Mixed storage — 53

3.3	Storage atmospheres	54
	3.3.1 Controlled atmosphere (CA) technology	55
	3.3.2 Atmosphere regeneration	56
	3.3.3 Modified atmosphere technology	59
3.4	Maintaining quality	60
	3.4.1 Disease	61
	3.4.2 Disorders	64
	3.4.3 Injury	66

4. Production of non-fermented fruit products   70
P. RUTLEDGE

4.1	Introduction	70
4.2	Fruit quality	70
4.3	Temperate fruit juices	72
	4.3.1 Orange juice	72
	4.3.2 Citrus juices	76
	4.3.3 Apple juice	76
	4.3.4 Pear juice	79
	4.3.5 Stone fruit juices	79
	4.3.6 Berry juices	80
4.4	Tropical fruit juices	80
	4.4.1 Pineapple juice	81
	4.4.2 Papaya purée	81
	4.4.3 Mango pulp	81
	4.4.4 Passionfruit juice	83
	4.4.5 Guava pulp	84
4.5	Clarification of fruit juices	84
4.6	Methods of preservation	86
	4.6.1 Thermal treatment of the juice	86
	4.6.2 Canning	87
	4.6.3 Aseptic processing	87
	4.6.4 Bottling	88
	4.6.5 Chemical preservatives	89
	4.6.6 Freezing	89
	4.6.7 Filtration sterilisation	89
4.7	Concentration of fruit juice	89
	4.7.1 Essence recovery	89
	4.7.2 Concentration	91
4.8	Products derived from fruit juice	92
	4.8.1 Fruit juice drink	93
	4.8.2 Fruit nectars	93
	4.8.3 Carbonated beverages	93
4.9	Adulteration of fruit juice	94

5. Cider, perry, fruit wines and other alcoholic fruit beverages   97
B. JARVIS

5.1	Introduction	97
5.2	Cider	98
	5.2.1 A brief history	98
	5.2.2 Cider and culinary apples	99
	5.2.3 Fermentation of cider	103
	5.2.4 Special types of cider	110
	5.2.5 The microbiology of apple juice and cider	111
	5.2.6 The chemistry of cider	118
5.3	Perry (poire)	120

5.4	Fruit wines	121
	5.4.1 Fruits used in fruit wine manufacture	122
	5.4.2 Processing of the fruit	123
	5.4.3 Fermentation of fruit wines	123
	5.4.4 Fruit pulp fermentations	126
	5.4.5 Alcohol-fortified wines	127
	5.4.6 Sparkling (carbonated) fruit wines	127
5.5	Fruit spirits and liqueurs	128
	5.5.1 Fruit spirits (*sic* fruit brandies)	128
	5.5.2 Apéritifs and liqueurs	129
5.6	Miscellany	131

6. Production of thermally processed and frozen fruit   135
G. BURROWS

6.1	Introduction	135
6.2	Raw materials	135
6.3	Canning of fruit	136
	6.3.1 Cannery hygiene	136
	6.3.2 Factory reception	137
	6.3.3 Peeling	137
	6.3.4 Blanching	138
	6.3.5 Choice of cans	138
	6.3.6 Filling	138
	6.3.7 Syrup	139
	6.3.8 Cut out	139
	6.3.9 Closing	140
	6.3.10 Exhausting	141
	6.3.11 Can vacuum	142
	6.3.12 Processing	142
	6.3.13 Finished pack pH values	145
6.4	Varities of fruit	145
	6.4.1 Apples	145
	6.4.2 Apricots	148
	6.4.3 Bilberries	148
	6.4.4 Blackberries	148
	6.4.5 Black currants	148
	6.4.6 Cherries	149
	6.4.7 Gooseberries	149
	6.4.8 Grapefruit	150
	6.4.9 Fruit salad	151
	6.4.10 Fruit cocktail	151
	6.4.11 Fruit pie fillings	152
	6.4.12 Loganberries	152
	6.4.13 Oranges	152
	6.4.14 Peaches	153
	6.4.15 Pears	154
	6.4.16 Pineapple	154
	6.4.17 Plums and damsons	155
	6.4.18 Prunes	155
	6.4.19 Raspberries	156
	6.4.20 Rhubarb	156
	6.4.21 Strawberries	157
6.5	Bottling	157
6.6	Freezing	158
	6.6.1 Freezing methods	159
	6.6.2 Storage	160

		6.6.3	Packaging	161
	6.7	Aseptic packaging		162
		6.7.1	Sterilisation/pasteurisation	162
		6.7.2	Packaging	163

7. The manufacture of preserves, flavourings and dried fruits — 165
R.W. BROOMFIELD

7.1	Preserves		165
	7.1.1	Ingredients	166
	7.1.2	Fruit for jam manufacture	166
	7.1.3	Other ingredients	168
	7.1.4	Product types and recipes	170
	7.1.5	Methods of manufacture	176
7.2	Fruits preserved by sugar: glacé fruits		183
7.3	Fruits preserved by drying		186
	7.3.1	Dried vine fruit production	186
	7.3.2	Dried tree fruit production	190
7.4	Flavourings from fruits		190
7.5	Tomato purée		190

8. The by-products of fruit processing — 196
R. COHN and A.L. COHN

8.1	Introduction		196
8.2	By-products of the citrus industry		196
	8.2.1	Citrus premium pulp (juice cells)	197
	8.2.2	Products prepared from peel and rag	198
	8.2.3	Citrus oils	203
	8.2.4	Comminuted juices	206
	8.2.5	Dried citrus peel	206
8.3	Natural colour extraction from fruit waste		208
	8.3.1	Extraction of colour from citrus peels	209
	8.3.2	Extraction of colour from grapes	209
8.4	Apple waste treatment		210
8.5	Production of pectin		211
	8.5.1	Characterisation of pectin and pectolytic enzymes in plants	211
	8.5.2	Pectin enzymes in plants used for production of pectin	212
	8.5.3	Commercial pectins and their production	212
	8.5.4	Different types of pectin and their application	217
	8.5.5	Application of pectin in medicine and nutrition	218

9. Water supplies, effluent disposal and other environmental considerations — 221
M.J.V. WAYMAN

9.1	Introduction		221
9.2	Water sourcing		221
9.3	Primary treatment		222
	9.3.1	Screening	222
	9.3.2	Colour removal	223
	9.3.3	Adjustment of pH	223
	9.3.4	Filtration	224
	9.3.5	Carbon adsorption	224
	9.3.6	Primary disinfection	225

9.4	Secondary treatment		226
	9.4.1 Boiler feedwater		226
	9.4.2 Cooling water		226
	9.4.3 Water for bottle-washing		226
	9.4.4 Water for fruit dressing		226
	9.4.5 Water for process use		227
	9.4.6 Water for special applications		227
	9.4.7 Cleaning-in-place (CIP)		228
9.5	Effluent planning		229
	9.5.1 Segregation		229
	9.5.2 Effluent transfer		230
	9.5.3 Effluent reception		231
	9.5.4 Treatment objectives		231
9.6	Effluent characterisation		232
	9.6.1 Suspended solids		232
	9.6.2 Oxygen demand		232
	9.6.3 Other parameters		233
	9.6.4 Effluent monitoring		233
9.7	Effluent treatment		233
	9.7.1 Solids removal		234
	9.7.2 pH adjustment		234
	9.7.3 Biological oxygen demand (BOD)		236
9.8	Forms of biological treatment		237
	9.8.1 The activated sludge process		237
	9.8.2 Percolating 'filters'		239
	9.8.3 High-rate filtration		239
	9.8.4 Mechanical contacting		240
9.9	Tertiary treatment		240
	9.9.1 Filtration		240
	9.9.2 Solids removal and disposal		240
	9.9.3 Recovery and re-use options		241
9.10	Environmental auditing		241
	9.10.1 Baseline assessment		241
	9.10.2 Period review		242
	9.10.3 Quality and environmental standards		242

Index   **244**

1 Introduction to fruit processing
R.B. TAYLOR

1.1 Processing on a global scale

World population continues to rise at a steady rate. United Nations sources forecast an estimated figure of over 6000 million at the millennium. Over half of this figure is represented by the population of Asia, and nearly three-quarters of the world population live in poor and developing countries. The advanced countries rely more and more upon imported goods and this has resulted in a requirement for products vulnerable to deterioration in storage to be harvested, processed, packed and sent around the globe, often in a matter of hours. This provides trade, employment and commercial markets on an international scale.

Before we explore the subject of fruit processing, it will be helpful to consider the focus of our labours, the fruit itself and its origins.

The Old Testament book of Genesis starts with its allegorical reference to the tree at the centre of the garden of Eden and its forbidden fruit. There is no other description of this given in the text (at least, not in the King James' version of the Bible), yet we all know that it was an 'apple' that Eve offered to Adam. In terms of an exact location of this revelation of the weaknesses of man, it has been postulated by theological scholars[1] that the garden of Eden existed as that lush and fertile ground situated between the Tigris and Euphrates rivers. There is evidence that the genus of fruit tree known as malus or apple may have originated in the same geographical region, in the area known today as Iraq. Whether coincidental, or not, this rather puzzling piece of information serves to illustrate that man has enjoyed a long-standing relationship with fruit.

During the twentieth century, advances in technology and social-economic factors, including population increases, have dictated the need for rapid communication, more effective transportation, and higher levels of efficiency in the production industries. This is often only answered by automation.

Industrialisation, beginning in the Western world during the eighteenth century, has provided the stimulus for both social and technological changes. Although it has brought with it many benefits, particularly for the most advanced nations, there is also a cost, because the purely commercial objectives frequently tend to influence ethical issues and there is a sense of being driven by our own success. It is against this background that the fruit processing industry finds itself.

Collectively, we are faced with perhaps the most tantalising of human challenges: how to feed the world! One cannot underestimate the importance of the contribution made by the fruit growing and processing industries. Fruit has long been valued as part of the staple diet of many living things; its presence is inexorably linked to the welfare of life on our planet.

The traditional cottage industries and farming practices of the fruit industry have been replaced by highly efficient agricultural techniques providing products of predetermined quality. The term 'factory farming', frequently used in reference to livestock, could be equally well applied to the fruit industry. Plant cultivars are produced and selected to provide maximum yields with the minimum of labour at harvest-time. The idyllic scene of a group of country folk picking fruit in an orchard, with ladders and hand-woven baskets, is now part of a bygone era. In many instances, it has been replaced by automated harvesting, where the fruit is mechanically separated from the branches. One such technique utilises a mechanical clamp which is applied to the tree-trunk and literally shakes the fruit into a collecting sheet held below on a circular frame that surrounds the base of the tree. Soft fruits, such as black currants and raspberries, are grown from specially cultured plant stock. Such fruit bushes have been pruned and trained against carefully spaced supports so that mechanical harvesters passing between the rows can collect the crop in a fraction of the time taken by the traditional method of hand-picking.

Harvested fruit is sorted, graded and stored. For soft fruits, long-term storage is preceded by rapid freezing at $-18°C$ to $-26°C$. Block-frozen and individually quick-frozen (IQF) fruit are held for processing. In the case of block-frozen fruit, the structure of the berries will have been greatly altered and such fruit will be suitable only for juice extraction or comminuting for purée. The IQF fruit will be suitable for whole fruit products, such as jams, conserves and yoghurts.

The facility to store fruit on a long-term basis enables us to plan and process outside, and in addition to, the normal harvesting periods.

1.2 Factors influencing processing

This book discusses the technical and scientific principles upon which the fruit processing industries depend.

At all stages of food production from the harvesting of the raw-material to the sale of the finished product, the industry is subject to legislative controls. These are concerned with handling hygiene, nutrition, label declarations, authenticity, additives, export and import tariff controls, packaging and inspection, to name but a few. Different countries will approach their legal responsibilities in different ways, and this is often a direct result of cultural differences. Anyone who is aware of the problems encountered within the European Union on the subject of harmonisation can be under no illusion as to the difficulties involved in reaching agreement on quite minor points of policy, let alone the major concerns. In the USA, the Food and Drug Administration enacts the rules to be passed by Congress and these will have to become law throughout the 50 States that comprise the union. In spite of the complexities, trade can still flourish on the international scene, through necessity.

The European and US trade blocs with their high levels of importation tend to set rigid rules concerning quality, for obvious reasons, and this has resulted in the producer countries adopting higher manufacturing standards than were previously encountered. International quality standards, such as EN 29000 (ISO 9000), provide a benchmark for the purchaser. Any new enterprise or a change of raw-material sourcing will require careful inspection of the manufacturer's operation before an agreement to proceed can be made.

1.3 Fruit types for processing

1.3.1 Pome fruits

The apple, as well as many other fruit types, is no longer restricted to its original indigenous area. Voyages of exploration during the Middle Ages and the opening up of trade routes were instrumental in bringing all manner of botanical specimens both back to Europe and to other continents, where careful husbandry led to the generation of new cultivars and hence to increases in trade. In the USA, there are records as early as 1647 of apples being grafted onto seedling rootstocks in Virginia, and by 1733 apples from American were finding their way into the London markets.

The 'Pome fruits', of which only the apple and pear are commercially important, are now grown in most temperate regions of the world. Argentina, Australia, Bulgaria, Canada, China, France, Germany, Hungary, Italy, Japan, Netherlands, New Zealand, Poland, South Africa, Spain, UK and the USA are perhaps the foremost of those countries growing apples on a considerable scale, for both home and export use.

Although the apple industry has developed into a major commercial force, it is interesting to note that world markets are dominated by perhaps no more than 20 dessert and culinary varieties, which have been selectively bred to display such characteristics as disease resistance, winter hardiness, appearance (colour and shape) and texture, coupled with high average yield (Figure 1.1b). Amongst these will be found Bramley's Seedling, Brayburn, Cox's Orange Pippin, Delicious, Golden Delicious, Discovery, Granny Smith, Jonathon and Newtown Pippin.

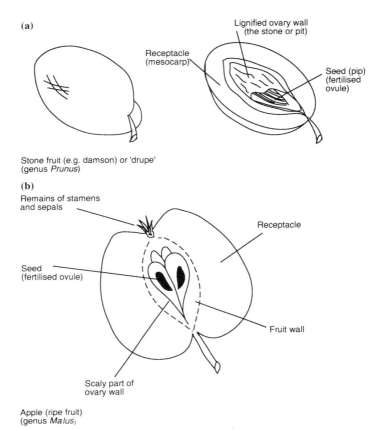

Figure 1.1 (a) A typical stone fruit or drupe, for example the damson (*Prunus*) and (b) structure of a ripe apple (*Malus*).

The main varities of pear of commercial importance are the Bartlett or Williams, Bon Chrétien, the Comice (Doyenné du Comice) and the Conference.

The majority of apples and pears are grown for direct consumption and, therefore, are graded at the point of harvest. Size, shape, colour and freedom from blemishes are of major importance to the retailer. Fruit not meeting such requirements, and yet quite sound, will be suitable for processing, where the selection criteria will relate to ripeness, flavour type (i.e. dessert or culinary), level of acidity, and soluble solids.

Fermentation processes such as those used for cider and perry manufacture may be classed as being traditional. Many of the more highly alcoholic beverages, such as French Calvados or the American apple brandy known as Applejack, are also traditional regional or national products. The ease with which apple juice tends to ferment has undoubtedly promoted its significance amongst the alcoholic beverages already mentioned, and a vast quantity of fruit is used annually in these products. With modern filtration and evaporation techniques, we are now able to produce aseptically packed concentrates that may be stored for extended periods in readiness for a variety of uses.

1.3.2 Citrus fruits

As with the pome fruits, citrus varieties are now grown in many parts of the world. Originating in the southern and eastern regions of Asia, China and Chochin China, and the Malayan Archipelago, the citron (*Citrus medica*) is said to have first found its way to Europe during the third century BC, when Alexander the Great conquered western Asia. Later, the orange and lemon were introduced to the Mediterranean regions when the Romans opened up navigable routes from the Red Sea to India. Since then, the cultivation of citrus fruit has spread from Europe to the USA, where notable growing areas are Florida and California, to South America, where Brazil has the largest share of the world's market for oranges and orange juice products, and to South Africa and parts of Australia. Existing markets in the Mediterrranean area include Israel, Sicily (lemons) and Spain.

Citrus processing industries, whilst primarily centred on juice production, provide many by-products. The citrus varieties have been described as berry fruits in which the hairs inside the ovary walls form juice sacs. These are contained within the characteristically highly coloured, oil-bearing peel (flavedo) or epicarp, and further encased in a pithy structure known as the albedo. During removal of the juice, modern processing techniques are able to carry out simultaneous separation of the oil by abrasion of the epicarp to rupture the oil cells. The oil is flushed off with water and recovered by centrifugation. The residual pithy material, peel and albedo, is also used in the production of citrus pectins, clouding agents for the beverage industry and in the preparation of comminuted citrus bases.

1.3.3 Stone fruits

The stone fruits are characterised by a fleshy mesocarp (pulp) surrounding a wood-like endocarp or stone, referred to in agricultural and processing circles as the 'pit' (Figure 1.1a). The epicarp or skin is thin and smooth, except in the case of the peach and apricot, which possess fine hair-like coatings. The stone fruit or commerce all belong to the genus *Prunus* (family Rosaceae) and include the species: peach, apricot, plum (and greengage), cherry and nectarine. Many varieties of plum occur and these are found in nearly all the temperature zones.

It is now generally thought that the peach and apricot were first grown and harvested in ancient China, although in relation to western civilisation many fruits originated in the area once known as northern Persia and the Russian provinces south of the Caucasus. Cherries and plums, however, were probably introduced into the western world sometime in advance of the apricot and peach. Wild cherries and plums would have spread rapidly because of the smaller size of fruit and the comparative ease of transfer by birds.

Unlike the majority of fruit types, the stone fruit is, in effect, a single seed or fertilised ovule surrounded by a fleshy receptacle, which, following the ripening stage, is valued for its texture and flavour. The seed and surrounding ovary are greatly enlarged in comparison with those of other fruits. The growth pattern of the fruit takes place in three stages. Initially, there is an enlargement of the ovary, the outer wall or shell of which remains soft and pliable. When an optimum size is attained, the shell begins to lignify, creating a wood-like protection about the seed and growth becomes confined to the endosperm and ovary. It is during the third phase that expansion of the mesocarp (edible portion) takes place and it is at this stage that there is a peak of biochemical activity. Increased sweetness, reduction in acidity and release of flavour components occur during ripening. This can take place before and after harvest, and with stone fruit care has to be taken to optimise this, as post-harvest ripening is generally thought to occur provided the fruit has reached the correct stage of maturity when picked. If harvested at too early a stage, the fruit will not reach the desired standard of maturity.

The feasibility of using a controlled atmosphere for stone fruit varieties was first mooted in 1967 as a means of extending storage time.[2] Controlled atmosphere storage, using low oxygen and/or carbon dioxide, is able to extend the 'life' of the fruit and this emphasises the 'living' nature of fruit cells. Different varieties of stone fruit show different responses to controlled atmosphere storage; this was noted by Couey,[3] who suggested an atmosphere of 7% O_2, 7% CO_2, and 86% N_2. This delayed ripening in plums and reduced the loss of soluble solids with no impairment of texture during storage at 0–1°C for 6 weeks.

Some plum varieties can be induced to ripen using ethylene treatment. Cherries are less responsive to controlled atmosphere storage and Porritt & Mason reported that different combinations of $O_2/CO_2/N_2$ were ineffective for storage of cherries.[4]

The ripening and biochemical activity experienced by many fruits are influenced by natural hormones.

1.3.4 Soft fruits

This term includes a number of unrelated fruits that have become grouped together in view of their size and culinary properties rather than for any structural or varietal reason. As might be expected by the term soft, these fruits do not store well. They are susceptible to moulds and yeasts and fruit held outside a comparatively short harvest season is usually stored deep frozen. The soft fruits are also all very prone to bruising at the time of harvest.

Three groups are found under this heading:

- the berry fruits (genus: *Rubus*): blackberry, raspberry, loganberry, boysenberry and mulberry (genus: *Morus*)
- the currants (genus: *Ribes*): gooseberry, currants (black, red and white) and blueberry (genus: *Vaccinium*)
- the achenes (false fruit): these are represented by the genus *Fragaria* (the strawberry family).

The structure of soft fruits. All these fruits fit the description of 'soft' (Figure 1.2). The berry fruits consist of aggregates of druplets surrounding a pithy receptacle rather than the situation occurring in the stone fruits where the whole fruit comprises just one druplet, or rather 'drupe' (originating from the Latin word *druppa* for an over-ripe olive), which is derived from an enlarged ovary containing a fertilised ovule or seed. The seeds of the currants are enclosed in a soft fleshy pericarp, and the number of seeds is characteristic of the fruit. In the gooseberry, the fruit may appear singly or in small clusters and can be picked at a single stage of maturity as required. The currants, however, produce their fruit on 'strigs' or short stems, with the fruit ripening in order along these stems commencing nearest the main branch and finishing with the terminal fruit. Therefore, even at the most favourable picking time, there will always be unripe and over-ripe fruit in the harvest. The achenes consist of a large pulpy, flavoursome receptacle supporting the seeds on the outer surface. There is no fleshy mesocarp surrounding the seeds.

1.4 Controlling factors in the ripening of fruit

1.4.1 Respiration climacteric

Respiration climacteric is a well-documented phenomenon characteristic of most but not all fruits during ripening. This stage of the growth phase of the fruit marks a sudden increase in metabolic activity. The term 'climacteric' rise was coined by Kidd and West to describe the sudden upsurge in the evolution of CO_2 that occurred during the ripening of apples (Figure 1.3).[5] Other fruits exhibit the phenomenon to a greater or lesser extent and considerable variation in pattern has been recorded. For the banana, preclimacteric respiration at 16–24°C can vary from 8–50 mg/h

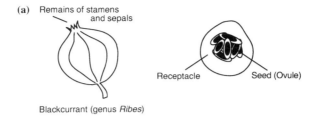

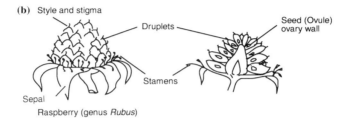

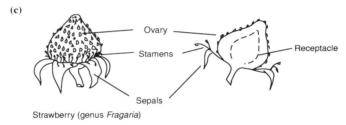

Figure 1.2 The structure of the soft fruits. Berry fruits: (a) currants, e.g. black currant (*Ribes*); (b) raspberry (*Rubus*); (c) achenes, e.g. strawberry (*Fragaria*).

CO_2 per kg fruit, whereas during the climacteric itself rates of 60 to about 250 mg/h CO_2 per kg fruit may be achieved.[6] As might be expected, the respiration rate will increase with increasing temperature.

By controlling the atmosphere surrounding fruits during storage, the ripening stage can be advanced or retarded; a factor of extreme importance in the marketing of fruit on a commercial scale.

1.4.2 Ethylene production

The gas ethylene is used to induce the climacteric phase in many fruits prepared for market. Kidd and West showed that the vapours produced by ripe apples when passed across unripe apples would stimulate respiration

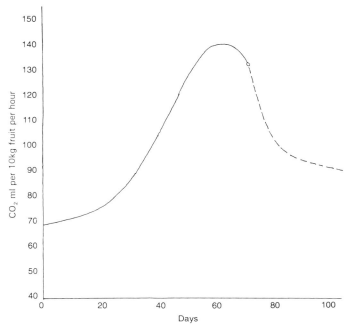

Figure 1.3 Respiration climacteric, characterised by the evolution of CO_2 during the ripening of apples.

into the climacteric phase and cause the unripe fruit to ripen.[7] Later it was shown that the active principle in the vapours was ethylene.[8]

Whilst the harvesting and marketing of any fruit crop is a critical operation, it is perhaps more so in the banana industry, where it is essential to control maturation of the fruit at all stages. Bananas are extremely susceptible to the effect of endogenous ethylene, and the picking of green banana fruit hastens ripening, apparently by lowering the theshold of sensitivity to ethylene.[9] To prevent this ripening effect during transportation and storage, the temperature is reduced to a level where ripening is not initiated by exposure to ethylene. The optimum storage temperature for bananas is about 13°C and, in practice, it is only necessary to protect the fruit from the effects of ethylene following harvest for 2–3 days until cooled to the optimum transportation temperature.

In addition to temperature control, the ripening effect can be delayed for extended periods by storing the green bananas in an atmosphere of 1–10% O_2, 5–10% CO_2 or a combination of low O_2 and high CO_2.[10,11] In this way, bananas can be stored for weeks or months whilst the onset of ripening is delayed.

Other fruits where controlled atmosphere storage is of major importance are pineapples, tomatoes and melons, and here the object is to meet the demands of the retail trade. Where used for processing purposes, the fruit is frequently sold in the form of pulp.

1.5 Biosynthesis of flavours

1.5.1 Analytical data

The flavour identity of each fruit variety plays an important part in its success on the commercial market, and whilst every effort is made during the farming of fruit to maintain and standardise flavour quality, the mechanisms and biological pathways by which specific flavouring components are formed are still largely unexplored.

Modern analytical methods enable the examination and identification of the chemical nature of each flavour, and for most fruit types there are extensive libraries, or lists, of chemicals that have been detected by mass spectral analysis. Perhaps the most widely used reference lists for *Volatile Compounds in Food* are those produced by the TNO–CIVO Food Analysis Institute of Holland, whilst many analytical laboratories and scientific organisations produce and maintain their own lists. TNO–CIVO have issued regular updates on their listings since 1963. Since then improvements in analytical techniques have resulted in the discovery of many compounds that occur at levels previously below the limit of detection. For example, the survey carried out in 1967 by Nursten and Williams reported on 137 compounds that had been identified in strawbery fruit aroma;[12] the TNO sixth edition report of 1989 lists no fewer than 351 volatile compounds, including 129 esters.[13]

Currently there are about 5000–6000 known flavouring components (not all of these isolated from fruit), whereas the flavourist may work from a list of no more than 500–600 in creating flavour matches of an acceptable standard. However, during the 1980s there has been a general requirement in the western world for 'naturalness' in flavourings and this has renewed interest in biological syntheses as well as in more effective methods of isolation of flavour components from natural sources.

1.5.2 Taste and aroma

The human organoleptic and gustatory senses will respond to remarkably low levels of stimuli. The flavour components responsible for characteristics in fruits may be present in extremely low concentrations, of the order of parts per million (p.p.m., 10^{-6}) or parts per billion (p.p.b., 10^{-9}), where the threshold of sensory detection may be achieved for many flavour chemicals. Most artificial flavourings are formulated to be used at a dose rate of about 0.1%. In the isolation of natural flavours from fruit sources by purely physical means, it is far more difficult to reach these levels of concentration, as interaction of components can occur during processing, resulting in degradation of flavour. Undoubtedly future development will take place to solve this problem within the constraints of manufacturing costs, but meanwhile for the majority of fruit aroma concentrates typical dosage rates are about 0.5%.

Whilst much is known about the identity of chemicals responsible for flavour, the pathways by which these chemicals have been formed have aroused less interest, and there is little widespread understanding as to how they are produced.

1.5.3 Flavour formation

In 1975, a systematic approach to the biosynthesis of flavours was made.[14] Five basic metabolic pools, covering the metabolism of carbohydrates, lipids, amino acids, terpenes and cinnamic acid, were cited, and examples were given of pathways leading to the formulation of mono- and sesquiterpenes, branched aliphatic esters, alcohols, acids, phenolic acids and ethers, and carbonyl compounds.

The flavour development period occurs during the climacteric rise in respiration; this is effectively the ripening stage (see Figure 1.4). Minute quantities of carbohydrates, lipids and protein and amino acids are converted to volatile flavours. The rate of flavour formation increases after the respiration climacteric and continues following the harvesting of the fruit up to the time that senescence sets in. Identification of the optimum stage of flavour development is often a highly subjective matter; for example, the European palate will often baulk at the sight of anything other than a yellow banana, whereas those originating from countries in which the banana is indigenous will require the skin of this fruit to be almost black before committing it to their taste-buds! Judging from the analytical flavour profile of 'overripe' bananas, the latter opinion must be correct (see Figure 1.5).

The chemicals responsible for characteristic aroma and flavour profiles of the different fruits may be grouped as follows.

Acids. Citric and malic acids are the most plentiful and widely dispersed of the non-volatile carboxylic acids and they are followed by tartaric, malonic, fumaric and gluconic (plums), ascorbic (vitamin C) and traces of benzoic, salicylic (bilberries), shikimic and quinic acids. Acids present in the volatile fractions of fruit are limited to two, namely, formic and acetic acids.

Carbonyls. Carbonyls make a significant contribution to the aroma and flavour of most fruits and are of major importance in certain instances: for

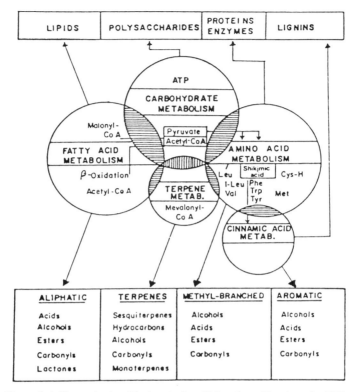

Figure 1.4 Biosynthesis of fruit volatiles. Taken from Tressl et al., 1975.

example, benzaldehyde in stone fruits, 5-hydroxy-2-methyl furfural in pineapples, acetaldehyde in oranges (i.e. orange essence oil) and furfural in strawberries.

Esters. These are the most important of the natural components that provide the flavour of fruits; they are highly variable and numerous and are responsible for the characterisation of fruit type.

Lactones. Lactones possess a functional carbonyl group and are often described as cyclic esters. They provide characteristic aromatic notes in many fruits, in particular peach and apricot.[15,16]

Phenols. Phenols are present in many fruits in the form of 'tannins'; these polymeric phenolic substances are associated with the sensation of astringency. They are largely undefined in terms of structure and are usually detected and measured by various colour reactions. 'Active' tannins decrease during ripening and in the ripe pulp have reduced to below 20% of their preclimacteric values.

1.5.4 Physiological and biochemical aspects

As the following chapters will indicate, fruit processing is very much a science-based operation and knowledge of the physiological and biochemical aspects of each fruit variety is advisable before subjecting it to any form of mechanical or enzymatic change. Industrial processing,

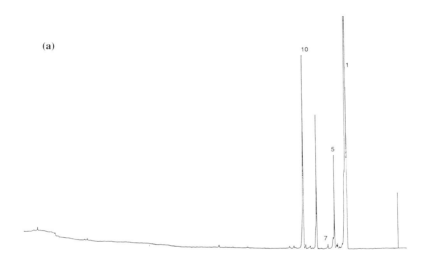

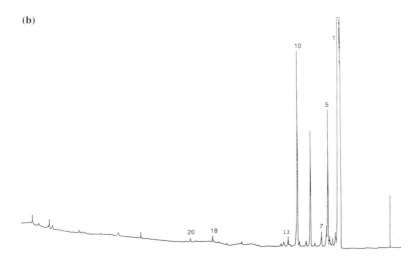

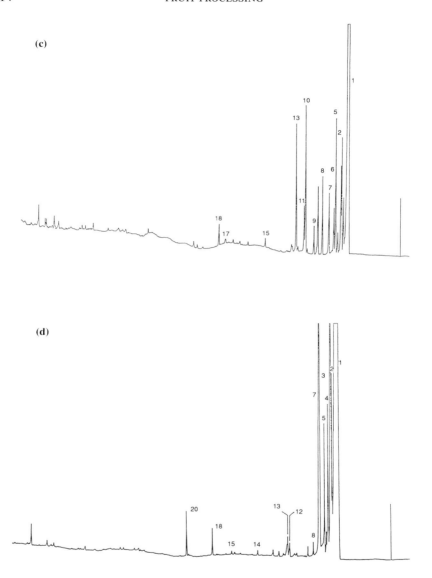

Figure 1.5 Peak identification for gas chromatograms showing profile changes in banana volatiles isolated from the fruit at various stages of maturity. (a) Underripe flesh; (b) ripe flesh; (c) well-ripened flesh; and (d) overripe flesh. Banana volatiles are labelled as follows: 1) ethanol; 2) ethyl acetate; 3) isobutanol; 4) *n*-butanol; 5) 3-methylbutan-1-ol; 6) 2-pentanol; 7) 3-methylbutanol; 8) isobutyl acetate; 9) butyl acetate; 10) *trans*-2-hexanol; 11) ethyl 2-methyl butyrate; 12) hexanol; 13) 3-methyl butylacetate; 14) 2-heptanol; 15) isobutyl-*n*-butyrate; 16) *n*-butyl-*n*-butyrate; 17) butyl-2-methylbutyrate; 18) 3-methylbutylbutyrate; 19) 3-methylbutyl-2-methylbutyrate; 20) 3-methylbutyl-3-methylbutyrate. Source: Research Department, Borthwicks Flavours Ltd, Wellingborough.

therefore, has become highly efficient and competitive in improving quality and presentation of the highly varied and extensive fruit product range.

The normal use of the term fruit refers to the receptacle, that is the casing around the true fruit or seed. The seed occurs as the result of sexual reproduction of the plant species concerned, and different functional components of the plant itself provide the protective environment or receptacle within which it can fully develop and eventually reproduce. For example, the edible part of an apple is the receptacle, which not only protects its fruit, the pips or seeds, but also provides an attractive packaging to animals to encourage ingestion and subsequent transportation of the seed to other locations.

Most small fruits (the word fruit is now being used in the more general sense) are readily transported and their seeds distributed via animal faeces. The eventual function of the receptacle in the larger fruits is to provide a bed of rotting humus in which their seeds can germinate.

The mechanism of attraction is a vital part of the reproductive cycle of plants and, just as the flower provides a perfume to lure certain insects into effecting pollination, so does the fruit itself, when mature, produce both aroma and flavour as an attractant to many animal and bird species. These organoleptic properties are also responsible for the commercial value placed upon fruit by humans.

1.6 Factors influencing fruit quality and crop yield

1.6.1 Fruit variety

Whilst each fruit type possesses its own characteristic flavour, there are varietal differences that have to be carefully assessed by the growers and processors to ensure success in the market place. The range of texture, of colour tones, of depth or freshness of flavour and the stability of these characteristics in any specific processing method to be employed are points to be considered by the processor when selecting his raw materials. The grower of fruit for the fresh market is primarily concerned by visual factors such as size, shape, colour and absence from blemishes and disease. The fruit processor, however, whilst showing preference towards high croppers on economic grounds, often has to apply higher and more restrictive standards of composition. Although the appearance of the fruit is still desirable for most processing needs, factors such as quality of flavour, texture, colour and nutritional value are going to be of major importance and, therefore, it is essential that the grower and processor maintain close liaison during the lead-up to harvest so that optimisation of yields can be effected.

For many years in the UK, strawberry cultivation was dominated by the variety Cambridge Favourite, a reasonably heavy cropper with good flavour. Also, the fruit was relatively easy to harvest, being readily detached from the plant. This particular variety is noted for its distinct lack of colour, the flesh of the fruit being white at the centre as the colour components are located on the inside of the fruit surface. Initially the poor colour had little or no effect on its use for jam production as strawberry is notorious for browning during the cooking process and 'brown tones' in the end-product were generally accepted as 'status quo'. Where appropriate and as the product demanded, the processor could always add colour either as natural anthocyanin extract or as one of more of the permitted artificial colourings (e.g. carmoisine and amaranth).

In recent years, new varieties of strawberry have appeared where, by careful selection and plant breeding, the colour has been enhanced while maintaining an acceptable flavour character. Varieties such as Totem and the more popular Senga Sengana provide an uninterrupted zone of colour across the entire fruit, and this combined with improvements in processing ensures a good colour in the finished product. This is particularly valuable in fruit juice concentrates, where a high quality standard is required. In addition, the use of artificial colourings is no longer an option where the product is to be declared 100% natural.

1.6.2 *External factors affecting fruit quality*

Other factors that affect fruit quality are geographical location of the growing area, climatic changes and the selection of the degree of maturity of the fruit at the point of harvest or, pertinently, at the point of processing. Storage characteristics of fruit vary with type; soft fruits, for instance, are extremely vulnerable to damage during handling and if not used within a few days of harvesting they need to be stored in deep-frozen conditions ($-18°C$ to $-26°C$). The pome fruits (e.g. apples and pears), however, are stored under controlled atmosphere, whereby maturation is effectively programmed to meet marketing requirements.

Storage conditions are understandably most critical as the maturation of the fruit is accompanied by changes in both the composition of flavour characteristics and the structural consistency or texture.

1.7 Flavour characteristics

There are four main groups contributing to the flavour characteristics in fruits. These are organic acids, sugars, bitter or astringent principles and volatile flavour constituents. The organic acids are usually citric, malic or lactic acids and these provide a tartness in flavour and give the thirst-quenching properties to soft drinks. Sugars contribute sweetness and body and astringency relates to certain phenolic compounds, e.g. tannins, saponins, naringin and hesperidin. These components are all present to a greater or lesser degree and constitute the background for perhaps the main characterisation factor, the aroma volatiles.

At the point of optimum flavour quality, occurring at the peak maturity, the fruit is highly vulnerable to deterioration. Processing is required to ensure preservation and/or presentation in a form suitable for marketing.

Processing will of necessity involve major structural changes in the fruit (e.g. slicing, pulping, pressing, heating and freezing) and this will be accompanied by a change in organoleptic characteristics as flavour compounds are affected *in situ*. Such changes may be beneficial, for instance in the manufacture of certain conserves where the 'jammy' flavour obtained by the extended cooking of some varieties of fruit may be preferred to the flavour achieved by a more natural approach where aroma volatiles normally lost during the initial cooking in the jam tun are replaced at a later stage in production to capture the whole fruit flavour.

1.8 The global market: threats and opportunities

Now that major fruit types (citrus, pome and berry fruits) can be grown in many parts of the world, it is perhaps not remarkable that a glut in production in a particular region can have far-reaching effects upon world prices, often sending them plummeting in an effort to remain competitive.

Under the Agriculture Policy of the European Union, it is the practice in such situations to adopt an interventionist approach by guaranteeing the producer a minimum price for his harvest, which is then effectively 'dumped' out of the market. In Europe there have been many instances of 'apple mountains' and 'wine lakes' as a result of overproduction.

By comparison, a dearth in production can lead to equally dire results, inviting underhand practices, whereby fruit pulps and juice concentrates are 'stretched' by the addition of sugars, acids and flavour top-notes, thus artificially increasing yields. Such practices became fairly commonplace in the early and mid-1980s and eventually resulted in the development of new analytical techniques to identify and combat production of highly sophisticated copies of the real thing. Perhaps the most effective of the new techniques is a method known as site-specific nuclear isotopic fraction by nuclear magnetic resonance (SNIF NMR) and this has been developed into a powerful analytical tool.

SNIF NMR represents the state of the art in the analysis of many naturally sourced products, including wines, spirits, fruit juices and flavours. It has rapidly become an important arbiter in the area of authenticity testing of products derived from, or said to be derived from, fruit sources. A typical scenario could involve the high value red fruit concentrates, e.g. strawberry, black currant, raspberry and blackberry. Concentrates are currently available on the market that have been produced by adulteration of the named source material with concentrates prepared from other berries, inferior quality fruit or even from the lower value deodorised and decolorised apple or pear juice concentrates. Adjustments are made by addition of citric (sometimes malic) acid and elderberry concentrate, or red grape extract is used to intensify colour. Depending upon the extent of the intended sophistication, aroma top-note chemicals are added to boost the flavour as deemed necessary. Whilst the finished blend may be natural, it is nevertheless non-authentic as regards the named fruit source. In the early 1980s, it would have been difficult to challenge 'natural copies' in terms of their authenticity, but nowadays it is beginning to be too costly to produce a 'copy' that can get past the analyst.

In conjunction with SNIF NMR techniques, high-performance liquid chromatography (HPLC) is used for anthocyanin determinations. Anthocyanin colours are widely distributed in nature and their distinctive profiles in many fruits allows their use as markets for the identification and classification of juices.

1.9 Fruit processing

This chapter has drawn attention to some of the factors to consider when working with fruit, and the following pages will frequently return to such parameters in regard to specific processing requirements. The processing operation, correctly designed, will need to succeed on three counts: (i) the plant should be adequately sized for the task and in good working order; (ii) the operator should be properly trained and following the correct instructions; and (iii) the raw-material fruit should be to the right specification. Whilst all three are essential, it is the last aspect that will always be of paramount importance.

References

1. E.W. Wade (1901). Old Testament History, Vol. 1. Methuen, p. 40.
2. R.E. Anderson, C.S. Parsons & W.L. Smith (1967). *Agric. Res. Wash.* **25**, 7.
3. P.M. Couey (1960). *Proc. Am. Soc. Hort. Sci.*, **75**, 207.
4. S.W. Porritt & J.L. Mason (1965). *Proc. Am. Soc. Hort, Sci.*, **87**, 128.
5. F. Kidd & C. West (1922). *Rep. Food Invest. Bd for 1921*, 17.
6. J.K. Palmer (1971). In *The Biochemistry of Fruits and their Products*, Vol. 2 (ed. A.C. Hulme).
7. F. Kidd & C. West (1933). *Rep. Food Invest. Bd for 1932*, 55.
8. R. Gane (1935). *J. Pomol.*, **13**, 351.
9. S.P. Burg & E.A. Burg (1965). *Bot. Gaz.*, **126**, 200.
10. R.E. Young, R.J. Romai & J.B. Biale (1962). *Plant Physiol.* (Lancaster), **37**, 416.
11. L.W. Mapson & J.E. Robinson (1966). *J. Food Technol.*, **1**, 215.

12. H.E. Nursten & A.A. Williams (1967). *Chem. Ind.*, 486.
13. H. Maarse & C.A. Visscher (ed.) (1989). *TNO–CIVO. Volatile Compounds in Food*, Vol 1 (16), p. 189.
14. R. Tressl, M. Holzer & M. Apetz (1975). In *Aroma Research* (ed. H. Maarse & P.J. Groenen), p. 41.
15. C.S. Tang & W.G. Jennings (1971). *J. Agric. Food Chem.*, **16**, 252.
16. R.J. Romani & W.G. Jennings (1971). In *The Biochemistry of Fruits and their Products*, Vol. 2 (ed. A.C. Hulme).

2 Fruit and human nutrition
P.C. FOURIE

2.1 Introduction

Fruits contain a wide range of different compounds and, therefore, show considerable variations in composition and structure. Each individual fruit is composed of living tissues that are metabolically active, and it is constantly changing in composition. The rate and extent of such changes depend on the physiological role and stage of maturity of the fruit concerned (Salunkhe, Bolin & Reddy, 1991). The nutritive value of fruit also depends on its composition. However, although fruit plays a very significant role in human nutrition, the composition of fruit is such that fruit is not recommended as a sole source of nutrition. However, it can be used advantageously to supplement deficiencies in other foods. In this chapter, the composition and importance of fruit in the diet, and the influence of processing on nutritional value of temperate fruits, citrus fruits and tomatoes are discussed. Tomatoes are included because they are botanically classified as a fruit, although they are marketed and consumed as a vegetable (Wills, Lim & Greenfield, 1984).

2.2 Composition of fruits

The most important components in fruit can be grouped as follows: water, proteins, carbohydrates, fats, minerals and vitamins. Most of these components are essential nutrients that are needed by the human body. The amount of each of these nutrients required by the body depends on factors such as age, mass, sex, health and physical activity. The requirements listed in nutritional tables merely indicate approximate daily requirements. Some of the nutrients are also required in very small quantities.

Water is the most abundant component of fruit (more than 80%), ranging from 82% in grapes to 90% in strawberries (Table 2.1) and even up to 93% in tomatoes (Table 2.2). The low value reported for figs (24%) is because in this specific case the product was semi-dried (Table 2.1). In normal circumstances, values up to 86% are reported for figs (Wills, Lim & Greenfield, 1987). However, the maximum water content varies between individual fruit of the same kind because of structural differences. It may

Table 2.1 Description and some compositional data on temperate fruits per 100 g

Fruit	Description	Edible proportion	Water (g)	Protein (g)	Carbo-hydrate (g)	Energy value (kcal)	Energy value (kJ)	Total nitrogen (g)	Starch (g)	Total sugars (g)	Dietary fibre Southgate method (g)	Dietary fibre Englyst method (g)
Apples	Flesh and skin	1.00	84.5	0.4	11.8	47	199	0.06	Tr	11.8	(2.0)	1.8
Apricots	Flesh and skin	1.00	87.2	0.9	7.2	31	134	0.14	0	7.2	1.9	1.7
Cherries	Flesh and skin	1.00	82.8	0.9	11.5	48	203	0.14	0	11.5	1.5	0.9
Figs	Semi-dried	1.00	23.6	3.3	48.6	209	889	0.52	0	48.6	11.4	6.9
Grapes	White, black and seedless	1.00	81.8	0.4	15.4	60	257	0.06	0	15.4	0.8	0.7
Nectarines	Flesh and skin	1.00	88.9	1.4	9.0	40	171	0.20	0	8.0	2.2	1.2
Peaches	Flesh and skin	1.00	88.9	1.0	7.6	33	142	0.16	0	7.6	2.3	1.5
Pears	Flesh and skin	1.00	83.8	0.3	10.0	40	169	0.05	0	10.0	N	2.2
Plums	Flesh and skin	1.00	83.9	0.6	8.8	36	155	0.09	0	8.8	2.3	1.6
Strawberries	Flesh and pips	0.95	89.5	0.8	6.0	27	113	0.13	0	6.0	2.0	1.1

Tr, trace; N, not determined.
Data from Holland et al., (1992) *The Composition of Foods*, 5th edn. are reproduced with the permission of The Royal Society of Chemistry and The Controller of Her Majesty's Stationery Office.

Table 2.2 Description and some compositional data on citrus fruit and tomatoes per 100 g

Fruit	Description	Edible proportion	Water (g)	Protein (g)	Carbo-hydrate (g)	Energy value (kcal)	Energy value (kJ)	Total nitrogen (g)	Starch (g)	Total sugars (g)	Dietary fibre Southgate method (g)	Dietary fibre Englyst method (g)
Grapefruit	Flesh only	1.00	89.0	0.8	6.8	30	126	0.13	0	6.8	(1.6)	1.3
Lemons	Whole, without pips	0.99	86.3	1.0	3.2	19	79	0.16	0	3.2	4.7	N
Oranges	Flesh only	1.00	86.1	1.1	8.5	37	158	0.18	0	8.5	1.8	1.7
Tangerines	Flesh only	1.00	86.7	0.9	8.0	35	147	0.14	0	8.0	1.7	1.3
Tomatoes	Flesh, skin and seeds	1.00	93.1	0.7	3.1	17	73	0.11	Tr	3.1	1.3	1.0

Tr, trace; N, not determined.
Data from Holland et al., (1992) *The Composition of Foods*, 5th edn. are reproduced with the permission of The Royal Society of Chemistry and The Controller of Her Majesty's Stationery Office.

also be affected by cultural conditions, which influence structural differentiation (Salunkhe et al., 1991).

Proteins usually contribute less than 1% of the fresh weight of fruit. Proteins are composed of amino acids, 10 of which are classified as essential for the human diet. These essential amino acids required by humans are valine, threonine, tryptophan, isoleucine, methionine, leucine, lysine, phenylalanine, histidine and arginine. They cannot be synthesized by the human body and must, therefore, be consumed regularly. A protein containing all ten essential amino acids is known as a complete protein (Potter, 1986). However, it is not sufficient for a complete protein merely to contain the essential amino acids, they must also be fully available to the body in the correct ratios. Incomplete proteins can be supplemented with the missing essential amino acids either in the form of synthetic compounds or as protein concentrates from natural sources. The protein content of fresh fruit is calculated by multiplying the total nitrogen content by a factor of 6.25. This figure is based on the fact that protein contains about 16% nitrogen and that all nitrogen, not considering other simple nitrogenous substances that may be present in the uncombined form, is present as protein. Simple nitrogeneous substances like asparagine and glutamine and their related acids, aspartic and glutamic acids, are abundant in citrus fruit, tomatoes and strawberries. The presence of asparagine is also very high in apples and pears; oranges are rich in proline (Salunkhe et al., 1991).

Carbohydrate consists of polysaccharides such as starch, cellulose, hemicellulose and pectic material and also of disaccharides and monosaccharides, such as the sugars sucrose, fructose and glucose. The amount of each of these constitutents can change drastically during ripening of fruit. Sugars are usually abundant when the fruit reaches its full maturity. In fruit containing starch, all the starch is then fully hydrolysed. These sugars are mostly glucose and fructose, but in fruit like peaches, nectarines and apricots the main sugar is sucrose (Wills, Scriven & Greenfield, 1983). Apples and pears are rich in fructose. Traces of other mono- and disaccharide sugars, such as xylose, arabinose, mannose, galactose and maltose, may also be present in fruit in small amounts. Sorbitol, a polyol related in structure to the sugars which is well known for its laxative effect, is present in relatively high concentrations in pears and plums, while no sorbitol has been reported in strawberries (Wrolstad & Shallenberger, 1981). Cellulose, hemicellulose and pectic material are the cell-wall components of fruit. Pectin can be used commercially for the manufacture of jams and jelly and can be extracted from the white spongy layer of citrus fruit skins, especially grapefruit and lemon, as well as from apples (see chapter 8). The total carbohydrate value reported in the tables of this chapter includes starch and total sugars. These values for total carbohydrate vary from 3% in lemons and tomatoes (Table 2.2) to about 15% in grapes (Table 2.1).

Dietary fibre, by definition, comprises the structural materials of plant cells that are resistant to the digestive enzymes of the stomach. It includes the structural polysaccharides of the cell wall and lignin (Ross, English & Perlmutter, 1985). The dietary fibre content of fresh fruit generally falls within the range 0.7–4.7% (Tables 2.1 and 2.2). Values quoted for dietary fibre in this chapter have been determined by the methods of Southgate and Englyst and also that by Cummings. The Southgate values are generally higher than the non-cellulosic polysaccharide values of Englyst and Cummings, because they include lignin. High dietary fibre values are usually associated with fruit of a lower water content or with fruit containing seeds in the edible portion. The fibre content can be reduced by removing the peel of the fruit. In apples, the reduction is approximately 11% and in pears it is approximately 34% (Jones et al., 1990).

The lipid content of fruit is generally below 1% and varies with the commodity (Tables 2.3 and 2.4). Apart from the fact that fats (oils) serve as sources of energy, the body also requires small quantities of unsaturated fatty acids. At least one of these, linoleic acid, is an essential fatty acid. However, fruit is not a good source of fat and of the temperate fruits, lemons and tomatoes have the highest fat content (both 0.3%) (Tables 2.3 and 2.4). The most important sources of energy in food are carbohydrates and fats (oils). In fruit, the most significant contribution to energy value is made by carbohydrates. Proteins and organic acids can also serve as a source of energy, but the body will preferably use carbohydrates and fats. The energy value is measured in heat units called calories and is expressed as kilocalories (kcal) or kilojoules (kJ) (kilocalories × 4.2 = kilojoules). In the temperate fruits, the energy values for tomatoes (73 kJ) and lemons (79 kJ) are the lowest (Table 2.2), while cherries (203 kJ) and grapes (257 kJ) are the best sources of energy (Table 2.1).

Fruit also contains a variety of essential mineral elements, although vegetables are far richer in minerals. About 14 mineral elements are considered to be essential nutritional constituents: calcium, sodium, zinc, iodine, copper, phosphorus, potassium, sulphur, fluoride, manganese, iron, magnesium, cobalt and chloride. Several other elements occur in low concentrations in human tissues and fluids but it has not yet been established that they are essential nutrients. Although fruits are not rich in minerals, potassium is the most abundant mineral available in fruit (Tables 2.5 and 2.6) and occurs mainly in combination with the various organic acids (Hugo, 1969). In fruit the pH of the tissue is controlled by the potassium/organic acid balance and high concentrations of potassium can contribute to blood pressure of humans. In strawberries and tomatoes, calcium, magnesium, phosphorus and chloride are the only minerals occurring in amounts exceeding 10 mg/100 g of fresh fruit tissue, but the mineral content usually varies considerably within fruit from a specific producing area. Calcium is always present in the pectic material in the cell

Table 2.3 The total fat, fatty acids and cholesterol composition of temperate fruits per 100 g

Fruit	Description	Fat (g)	Fatty acids			Cholesterol (mg)
			Saturated (g)	Monounsaturated (g)	Polyunsaturated (g)	
Apples	Flesh and skin	0.1	Tr	Tr	0.1	0
Apricots	Flesh and skin	0.1	Tr	Tr	Tr	0
Cherries	Flesh and skin	0.1	Tr	Tr	Tr	0
Figs	Semi-dried	1.5	N	N	N	0
Grapes	White, black and seedless	0.1	Tr	Tr	Tr	0
Nectarines	Flesh and skin	0.1	Tr	Tr	Tr	0
Peaches	Flesh and skin	0.1	Tr	Tr	Tr	0
Pears	Flesh and skin	0.1	Tr	Tr	Tr	0
Plums	Flesh and skin	0.1	Tr	Tr	Tr	0
Strawberries	Flesh and pips	0.1	Tr	Tr	Tr	0

Tr, trace; N, not determined.
Data from Holland et al., (1992) *The Composition of Foods*, 5th edn. are reproduced with the permission of The Royal Society of Chemistry and The Controller of Her Majesty's Stationery Office.

Table 2.4 The total fat, fatty acids and cholesterol composition of citrus fruits and tomatoes per 100 g

Fruit	Description	Fat (g)	Fatty acids			Cholesterol (mg)
			Saturated (g)	Monounsaturated (g)	Polyunsaturated (g)	
Grapefruit	Flesh only	0.1	Tr	Tr	Tr	0
Lemons	Whole, without pips	0.3	0.1	Tr	0.1	0
Oranges	Flesh only	0.1	Tr	Tr	Tr	0
Tangerines	Flesh only	0.1	Tr	Tr	Tr	0
Tomatoes	Flesh, skin and seeds	0.3	0.1	0.1	0.2	0

Tr, trace.
Data from Holland *et al.*, (1992) *The Composition of Foods*, 5th edn. are reproduced with the permission of The Royal Society of Chemistry and The Controller of Her Majesty's Stationery Office.

Table 2.5 The inorganic constituents of temperate fruits per 100 g

Fruit	Description	Na (mg)	K (mg)	Ca (mg)	Mg (mg)	P (mg)	Fe (mg)	Cu (mg)	Zn (mg)	Cl (mg)	Mn (mg)	Se (µg)	I (µg)
Apples	Flesh and skin	3	120	4	5	11	0.1	0.02	0.1	Tr	0.1	Tr	Tr
Apricots	Flesh and skin	2	270	15	11	20	0.5	0.06	0.1	3	0.1	(1)	N
Cherries	Flesh and skin	1	210	13	10	21	0.2	0.07	0.1	Tr	0.1	(1)	Tr
Figs	Semi-dried	62	970	250	80	89	4.2	0.30	0.7	170	0.5	Tr	N
Grapes	White, black and seedless	2	210	13	7	18	0.3	0.12	0.1	Tr	0.1	(1)	1
Nectarines	Flesh and skin	1	170	7	10	22	0.4	0.06	0.1	5	0.1	(1)	3
Peaches	Flesh and skin	1	160	7	9	22	0.4	0.06	0.1	Tr	0.1	(1)	3
Pears	Flesh and skin	3	150	11	7	13	0.2	0.06	0.1	1	Tr	Tr	1
Plums	Flesh and skin	2	240	13	8	23	0.4	0.10	0.1	Tr	0.1	Tr	Tr
Strawberries	Flesh and pips	6	160	16	10	24	0.4	0.07	0.1	18	0.3	Tr	9

Tr, trace.
Data from Holland *et al.*, (1992) *The Composition of Foods*, 5th edn. are reproduced with the permission of The Royal Society of Chemistry and The Controller of Her Majesty's Stationery Office.

Table 2.6 The inorganic constituents of citrus fruits and tomatoes per 100 g

Fruit	Description	Na (mg)	K (mg)	Ca (mg)	Mg (mg)	P (mg)	Fe (mg)	Cu (mg)	Zn (mg)	Cl (mg)	Mn (mg)	Se (µg)	I (µg)
Grapefruits	Flesh only	3	200	23	9	20	0.1	0.02	Tr	3	Tr	(1)	N
Lemons	Whole, without pips	5	150	85	12	18	0.5	0.26	0.1	5	N	(1)	N
Oranges	Flesh only	5	150	47	10	21	0.1	0.05	0.1	3	Tr	(1)	2
Tangerines	Flesh only	1	120	31	8	12	0.2	0.01	0.1	1	Tr	N	N
Tomatoes	Flesh, skin and seeds	9	250	7	7	24	0.5	0.01	0.1	55	0.1	Tr	2

Tr, trace; N, not determined.
Data from Holland et al., (1992) *The Composition of Foods*, 5th edn. are reproduced with the permission of The Royal Society of Chemistry and The Controller of Her Majesty's Stationery Office.

walls of the fruit, magnesium in the chlorophyll molecules and phosphorus can play an important part in carbohydrate metabolism. In general, mineral elements widely contribute to the quality of fruit products. Calcium, for example, can influence the texture and storage life of fruit.

Vitamins are nutrients required for specific functions in the body. If the vitamins are not consumed in sufficient quantities, deficiency diseases develop. However, in most instances very small quantities are required. Considerable differences in vitamin content are reported between fruit species and varieties as well as between the same variety grown under different environmental conditions. Climate, soil and fertilizer practices also affect the levels of vitamins in fruits. Fruit is especially known as a source of ascorbic acid, although tropical fruits are a better source than the similar products grown in temperate regions. Strawberries and black currants are good sources of ascorbic acid, citrus fruits contain moderate amounts, while apples, pears, cherries and plums contain very little (Tables 2.7 and 2.8). An important environmental factor controlling the level of ascorbic acid is sunlight. Generally, the greater the amount of sunlight during growth, the greater the ascorbic acid content (Salunkhe *et al.*, 1991). Tomatoes with heavier shading have been shown to have lower ascorbic acid levels than those with more light exposure (Bradley, 1972). Fruit possesses an advantage over many vegetables because ascorbic acid is present in the acidic environment (where it is more stable) provided by fruit juices compared with the more neutral environment in vegetables. A further advantage is that fruit is eaten raw and the possible loss of ascorbic acid through cooking is prevented (Bradley, 1972)

Vitamin A is fat soluble and does not occur as such in fruit, although certain fruit carotenoids can be converted to vitamin A in the body. These pigments are referred to as provitamin A. Most of the provitamin A in fruit is in beta-carotene, with lesser amounts in alpha-carotene, gamma-carotene and other carotenoid pigments (Bradley, 1972). Fruits, in general, are not good sources of carotene, but in temperate regions apricots and plums are moderate sources, while peaches, nectarines and tangerines contain only small amounts. Lycopene represents 90% of the total carotenoid fraction in tomatoes, but does not yield vitamin A potential (Tables 2.7 and 2.8).

Fruit is a moderate to poor source of the members of the vitamin B group. The average requirement by the human body of these vitamins is also low, and the amounts available in fruits are usually sufficient for human nutrition. Plums (Table 2.7) and tomatoes (Table 2.8) are a good source of niacin, and strawberries (Table 2.7), oranges and grapefruit (Table 2.8) have significant amounts of folate. Vitamin B_{12} is the only vitamin in the B group that is not present in fruit. The rest of the vitamins, like vitamin D, which also is a fat-soluble vitamin, is also absent in fruit,

Table 2.7 The vitamin content of temperate fruits per 100 g

Fruit	Description	Retinol (μg)	Carotene (μg)	Vitamin D (μg)	Vitamin E (mg)	Thiamin (mg)	Ribo-flavin (mg)	Niacin (mg)	Trypt-ophan 60 (mg)	Vitamin B₆ (mg)	Vitamin B₁₂ (μg)	Folate (μg)	Panto-thenate (mg)	Biotin (μg)	Vitamin C (mg)
Apples	Flesh and skin	0	18	0	0.59	0.03	0.02	0.1	0.1	0.06	0	1	Tr	1.2	6
Apricots	Flesh and skin	0	405	0	N	0.04	0.05	0.5	0.1	0.08	0	5	0.24	N	6
Cherries	Flesh and skin	0	25	0	0.13	0.03	0.03	0.2	0.1	0.05	0	5	0.26	0.4	11
Figs	Semi-dried	0	(59)	0	N	0.07	0.09	0.7	0.4	0.24	0	8	0.47	N	1
Grapes	White, black and seedless	0	17	0	Tr	0.05	0.01	0.2	Tr	0.10	0	2	0.5	0.3	3
Nectarines	Flesh and skin	0	58	0	N	0.02	0.04	0.6	0.3	0.03	0	Tr	0.16	(0.2)	37
Peaches	Flesh and skin	0	58	0	N	0.02	0.04	0.6	0.2	0.02	0	3	0.17	(0.2)	31
Pears	Flesh and skin	0	18	0	0.50	0.02	0.03	0.2	Tr	0.02	0	2	0.07	0.2	6
Plums	Flesh and skin	0	295	0	0.61	0.05	0.03	1.1	0.1	0.05	0	3	0.15	Tr	4
Strawberries	Flesh and pips	0	8	0	0.20	0.03	0.03	0.6	0.1	0.06	0	20	0.34	1.1	77

Tr, trace; N, not determined.
Data from Holland et al., (1992) *The Composition of Foods*, 5th edn. are reproduced with the permission of The Royal Society of Chemistry and The Controller of Her Majesty's Stationery Office.

Table 2.8 The vitamin content of citrus fruits and tomatoes per 100 g

Fruit	Description	Retinol (µg)	Carotene (µg)	Vitamin D (µg)	Vitamin E (mg)	Thiamin (mg)	Ribo-flavin (mg)	Niacin (mg)	Trypt-ophan 60 (mg)	Vitamin B$_6$ (mg)	Vitamin B$_{12}$ (µg)	Folate (µg)	Panto-thenate (mg)	Biotin (µg)	Vitamin C (mg)
Grapefruit	Flesh only	0	17	0	(0.19)	0.05	0.02	0.3	0.1	0.03	0	26	0.28	(1.0)	36
Lemons	Whole, without pips	0	18	0	N	0.05	0.04	0.2	0.1	0.11	0	N	0.23	0.5	58
Oranges	Flesh only	0	28	0	0.24	0.11	0.04	0.4	0.1	0.10	0	31	0.37	1.0	54
Tangerines	Flesh only	0	71	0	N	0.05	0.01	0.1	0.1	0.05	0	15	0.15	N	22
Tomatoes	Flesh, skin and seeds	0	640	0	1.22	0.09	0.01	1.0	0.1	0.14	0	17	0.25	1.5	17

Tr, trace; N, not determined.
Data from Holland et al., (1992) *The Composition of Foods*, 5th edn. are reproduced with the permission of The Royal Society of Chemistry and The Controller of Her Majesty's Stationery Office.

while vitamin E only occurs in small quantities in some fruits (Tables 2.7 and 2.8).

The remainder of the nutrients are minor compositional components, but they also play an important role in the appearance, colour, taste and aroma of fruit. Organic acids play an important role in fruit taste through the sugar/acid ratio. Sugar provides sweetness and the organic acids sourness. The main organic acids present in fruit are citric and malic acids. Citrus fruit, strawberries, pears and tomatoes are dominated by citric acid while apples, plums, cherries and apricots produce primarily malic acid. In peaches, the two acids are in equal amounts. Grapes differ from other fruit in that tartaric acid is the main acid present. In all these fruits, there is a considerable decrease in acidity during ripening of the fruit. Bitterness is not common in fruit, but certain flavonoids, such as naringin in grapefruit and limonene in citrus fruit, are intensely bitter. The aroma of fruit is a key factor in assessing quality as well as identity (Salunkhe *et al.*, 1991). The major chemical compounds associated with aroma are esters of aliphatic alcohols and short-chain fatty acids. However, except for some fruits in which the major volatile compound is isoamyl acetate, there is still uncertainty as to the chemical substance responsible for the specific odour of most fruit.

Chlorophylls, carotenoids and anthocyanins are pigments that are responsible for the colour of fruit. Chlorophyll provides the green colour of, for example, apples, and carotenoids, mainly beta-carotene and lycopene, the yellow to orange colour of citrus fruit, apricots and peaches and the red colour of tomatoes. Anthocyanins provide the naturally red, blue or purple colour of apples, plums, grapes and strawberries.

Enzymes are important in fruit because of the chemical changes that they initiate. Ficin is a proteolytic enzyme only occurring in figs. This enzyme is present in the latex of the figs and reacts with proteins of the human skin when the figs are handled. This reaction causes dermatitis, and some individuals are so adversely affected that they cannot ingest fresh figs (Macrae, Robinson & Sadler, 1993). Phenoloxidases in apples, pears, grapes, strawberries and figs are responsible for the discoloration of cut surfaces when exposed to air. Citrus fruit and tomatoes are rich in pectin esterase, and pears and tomatoes in polygalacturonase, both being pectolytic enzymes responsible for softening during ripening.

2.3 The importance of fruit in the human diet

Nutrients frequently consumed in sub-optimal concentrations by humans are proteins, calcium, iron, vitamin A, thiamine (vitamin B_1), riboflavin (vitamin B_2) and ascorbic acid. Members of this group are often known as the critical nutrients. Fruit also contains these critical nutrients, and sometimes some of these nutrients occur in higher concentrations than in other foods. Therefore, fruit plays an important role in balancing the human diet, mainly because the composition of fruit differs markedly from other foods of plant and animal origin (Hugo, 1969).

The contribution of fruit to the protein requirement is slight and they are also not good sources of calcium, thiamine and riboflavin. However, some fruit is particularly rich in iron, carotene and ascorbic acid and lack of fruit in the diet will lead to a deficit in the vitamin C and beta-carotene content (Walter, 1994). Fruit with such a composition combined with low-calorie values is frequently employed by dieticians in drawing up low-calorie diets.

In experiments on nutrition, evidence has often been found indicating that ascorbic acid of natural origin is apparently superior to the synthetic product. It was established later that this phenomenon is caused by the presence of certain flavonoid compounds in fruit that influence the blood circulation, increasing the permeability and the elasticity of the capillary vessels. This action is known as vitamin P activity. These flavonoid substances are not classified as vitamins because several compounds have this property and no severe deficiency diseases develop if these substances are not present in food. There are indications that these flavonoids have a protective action against infections of the respiratory system. Unfortunately, they are readily decomposed in the body so that it is impossible to maintain high enough concentrations in the blood.

There are considerable differences of opinion as to the minimum daily requirement of ascorbic acid. In some countries, a daily allowance of 20 mg is considered sufficient, but in other countries quantities of up to 70 mg a day are recommended. Portions of 250 g strawberries, oranges and grapefruit contain considerably more than the minimum daily requirements of ascorbic acid, while most of the other fruit can provide more than half of our daily requirement. Vitamin C is also quite often added to fruit juices in addition to the natural level of this vitamin (Walter, 1994). Deficiencies in vitamin C tend to occur in temperate regions during the winter months and in tropical regions during periods of drought (Salunkhe et al., 1991).

Portions of 250 g apricots should provide sufficient vitamin A to satisfy the daily requirements. Peaches contain approximately 50% of the daily vitamin A requirements and 250 g portions of plums, strawberries, grapes and oranges contain more than 10% of the daily requirements. The bioavailability of certain other vitamins can differ greatly according to the foodstuff. The fat-soluble vitamins like vitamin A are only absorbed in significant amounts if the food consumed also contains lipids (Walter, 1994).

In the mature fruit stage, the organic acids and sugar occur in fruit in characteristic ratios. In most cases, the pH is lower than 4.2. Ripe fruit are higher in sugar content, but processors prefer unripe or partially ripened fruit as their juice is extracted more readily (Akhavan & Wrolstad, 1980).

However, when fresh, matured fruit is eaten, a mixture of natural sugars and not a single sugar is consumed (Koeppen, 1974). As a result, fruit and fruit beverages are particularly appetizing. Although fruits contain fructose, glucose and sucrose in various combinations, those containing high concentrations of free sugars should be restricted in the diet of diabetic patients (Nahar, Rahman & Mosihuzzaman, 1990). The response of blood glucose to fructose is low, while the response to sucrose is intermediate between that to fructose and that to glucose (Miller, Colagiuri & Brand, 1986). However, oral fructose can be insulinogenic in humans when blood glucose levels are elevated (Reiser *et al.*, 1987). Therefore, dieticians need more detailed information regarding the content of individual sugars in fruits when planning special diets for diabetics.

The minerals present in fruit are basic by nature, with the result that the residues of metabolism are alkaline. In comparison with other foods, fruit has a relatively high potassium to sodium ratio. Though some fruits, such as apricots, pears, peaches, plums and strawberries, contain oxalic acid, the concentration of this acid is not sufficient to reduce the absorption of calcium from the fruit or from other foods eaten together with the fruit. Acids occurring in fruit are almost completely metabolized in the body. From experiments carried out with animals, it has been established that acids from fruit favour the development of dental caries. In rats fed on different types of fruit juice, a relationship between the acidity of the juice and the development of caries was established. In humans this phenomenon has not yet been studied thoroughly. It has, however, been established in experiments carried out in different countries that children eating apples regularly were much less prone to dental caries and their gums were healthier than those children who did not eat apples regularly. The texture of apples is such that it has a cleansing action on the teeth.

The presence of about 10% (on a dry mass basis) of indigestible material or crude fibre in food is required for normal digestion. In human nutrition, this fact is often overlooked. Fruit contains 0.3–5.7% (2.6–33% on a dry mass basis) indigestible material, consisting of cellulose and related substances. Inclusion of fruit in the diet is, therefore, essential for supplying dietary fibre. It provides an indigestible matrix that stimulates the activity of the intestines and helps to keep the intestinal muscles in working order. Chronic constipation is probably largely alleviated by an increased intake of fresh fruits. However, much of the fibrous materials of stone and pome fruits are lost when these fruits are peeled. Whatever the mechanism, dietary fibre does play a part in a reduction of cholesterol levels in blood. However, all sources of dietary fibre do not have the same effects on blood serum cholesterol levels (Salunkhe *et al.*, 1991). In recent years, lack of fibre in the diet has been shown to cause colon cancer in humans (Ross *et al.*, 1985).

Fruit and fruit beverages can also be used advantageously as snacks (especially with children) instead of sweets and synthetic beverages. Fruit eaten as snacks should not adversely affect food intake during regular meals, because fruit is digested relatively quickly. Because of their ease of digestion, high vitamin content and appetizing effects, fruit juices are particularly useful to the indisposed. Obviously, the nature of the illness should be taken into account. In the treatment of obesity, fruit and fruit juice can be included advantageously in the diet in a regimen where only fruit and fruit juice are consumed on certain days of the week. On such days, the calorie intake is low while the high potassium-to-sodium ratio assists the excretion of excessive salt and water. According to dieticians, the consumption of fruit stimulates the excretion of water, with the result that the functions of the heart are eased. For this reason, fruit and fruit juices are also indispensible in the treatment of dropsy, chronic heart diseases and poor blood circulation. Fruit and fruit juices are also useful in the diet of patients suffering from kidney complaints, gastritis, gall and liver complaints and high fever.

Many therapeutic drugs in use in modern medicine originated as plant extracts. Therefore, it is not surprising that certain fruit components exert pharmacological or therapeutic effects. Limonin and nomilin (and other limonoids) are present in citrus fruit such as oranges, lemons and grapefruit. These compounds are believed to have a role in inhibiting the development of certain forms of cancer. Prunes contain hydroxyphenylisatin derivatives, which stimulate colonic smooth muscles, explaining their traditional use as a laxative (Macrae *et al.*, 1993). Sorbitol in pears and plums also serves as a laxative.

Several components with antioxidant activity naturally occur in fruit. These components include ascorbic acid, tocopherols (vitamin E), beta-carotene and other flavonoid components. The antioxidant activity of tocopherols is believed to be the main source of their vitamin activity. Vitamin C has specific physiological functions that are not dependent on its antioxidant properties (Macrae *et al.*, 1993). Beta-carotene has antioxidant properties that help neutralize free radicals, which are reactive and highly energized molecules formed through normal biochemical reactions or through external factors such as air pollution or cigarette smoke. Beta-carotene is mainly active as a quencher of singlet oxygen, which can induce precancerous changes in cells.

2.4 Changes in nutritive value during processing

Probably the single most important factor in the quality of fruit is the maturity of the fruit at harvest. It is important to distinguish between the stage that represents optimal eating quality and the stage of full biological maturation. In certain fruits, such as oranges, the two stages coincide. In fruit such as peaches, ascorbic acid decreases with advanced maturity. Ascorbic acid synthesis also continues during ripening of detached tomatoes, although the ascorbic acid content in a ripe tomato tends to be higher in fruit that is harvested at a later stage of maturity. Carotenoids tend to increase during ripening of fruits such as peaches and pears, although much of the green to yellow colour change is the result of unmasking of carotenoids through a loss of chlorophyll. Major changes have also been observed in the mineral content of grapes during maturation on the vine (Shewfelt, 1990). Contents of total sugar increase with ripening especially in cherries and plums (Hulme, 1971). Any type of harvesting technique that results in bruising generally leads to an increase in degradative reactions and a loss of nutritional quality, although the extent of nutrient loss varies widely between fruit species (Shewfelt, 1990).

Preservation is a convenient method of storing fruit for use in periods when the fresh products are not available. The characteristics of fruit are usually altered to such an extent during processing that the processed products do not necessarily resemble the fresh products. Some fruits, such as clingstone peaches and sultana (grapes), are grown almost entirely for canning and drying, respectively. If processed and stored properly, the nutritive value of these processed fruits is comparable to that of the fresh products.

Apart from the effect of the addition of sugar, water and some other ingredients, usually only slight changes occur in the nutritive value of fruit during processing. In the case of dried fruit and fruit juice concentrates, the evaporation of water results in a concentration of the fruit constituents. Dietary fibre content will be affected by processing methods such as peeling, juice extraction and drying as well as by cultivar and variation in horticultural practices (Jones *et al.*, 1990).

Application of poor processing techniques to fruit may result in considerable losses of ascorbic acid. This vitamin is readily destroyed by oxidation, especially in the presence of certain enzyme systems and metal ions. In sound fruit, ascorbic acid is not subjected to enzymatic oxidation. When fruit is peeled or macerated, however, this vitamin is subjected to oxidation because of the presence of gaseous oxygen and oxidative enzymes. In deciduous fruits, the enzymatic change is characterized by browning of the fruit tissue. The appearance of a brown colour is usually an indication that practically all of the ascorbic acid has been destroyed in that part of the fruit. Ascorbic acid is also rapidly destroyed when exposed to heat, especially in the presence of light or air and at neutral pH (6–7). It is relatively stable in more acidic foods. This is an important advantage, because acidic products such as tomato and citrus juices are important sources of this vitamin (Institute of Food Technologists, 1986). This means that during canning of fruit a relatively small portion of the ascorbic acid is destroyed, with the result that most canned fruits are usually good sources of this vitamin. Canned fruit often contains more ascorbic acid than the corresponding fresh fruit stored under unfavourable conditions. Nutrient losses caused by canning depend on factors such as type of fruit, type of container and the severity of the heat process. Minimum nutrient losses will occur if proper processing and cooking techniques are used. The processing methods for tomato sauce and tomato juice can cause an ascorbic acid loss of up to 34% (Farrow, Kemper & Chin, 1979). Loss of vitamins also occurs in some canned fruit products after processing. The loss of ascorbic acid depends mainly on storage conditions. Canned fruit can be stored at room temperature for long periods without appreciable changes. This, of course, is one of the advantages of such products.

Most fruit cans are exhausted prior to canning. The fruit is filled in a container and it usually passes through steam before sealing to reduce oxygen in the head space and prevent oxidative changes and off-flavours. Blanching may also be applied before freezing, juicing or, in some cases, dehydration of fruit. The fruit may be blanched by exposure to either near boiling water, steam or hot air for 1 to 10 minutes. Blanching inactivates enzyme systems that degrade flavour and colour and that cause vitamin losses during subsequent processing and storage. Blanching also removes air from the fruit tissues, destroys some of the contaminating microorganisms present and causes softer texture. Nutrient losses caused by blanching result directly from leaching of water-soluble vitamins into the water used during processing. Blanching with steam, hot air or microwaves does not require immersion in water and reduces leaching of vitamins (Institute of Food Technologists, 1986).

Fruit may be dried by sun drying or various evaporation processes. During dehydration, losses of vitamin C may vary from 10% to 50% and that of vitamin A from 10% to 20% (Institute of Food Technologists, 1986). The addition of sulphur dioxide during the drying of fruit has been claimed to have a beneficial effect on the retention of ascorbic acid and carotene, since it inhibits oxidation and it also prevents enzymatic browning. Thiamine, however, is destroyed by the presence of sulphur dioxide. Since the thiamine content of fruit is usually low, more value is attached to the retention of ascorbic acid and carotene. Despite the long history of utilization of sulphur dioxide, there is today an awareness of potential hazards associated with exposure to sulphur dioxide gas and several cases of respiratory diseases and atmospheric pollution have been reported (Wedzicha, 1984). The amount of sulphur dioxide permitted in dried fruit is controlled through regulations and varies from country to country. Sulphur dioxide levels in dried fruit also decrease during storage.

The nutritive value of fruit beverages depends mainly on the type of fruit used, methods of processing and the degree of dilution. As in canned fruit, the vitamin content of fruit beverage is lower than that of the original fruit.

However, vitamin C loss is greater in orange juice than in grapefruit juice at the same storage temperature as a result of non-enzymic aerobic and anaerobic reactions. In the preparation of fruit nectars, only part of the fibrous material is removed. Clear beverages, however, contain no fibres. During clarification, the pectic substances are decomposed with the aid of pectolytic enzymes (Hugo, 1969). Methyl alcohol is produced, but in all experiments carried out so far it has been found that the concentration of this alcohol is so low that no health hazard exists. Fruit juices are also packed aseptically with high-temperature/short-time techniques and this provides increased nutrient retention and fruit juice quality (Institute of Food Technologists 1986).

In the freezing of strawberries and other berries, no pretreatments are applied and practically no changes in nutritive value occur during processing and storage if proper packaging and processing conditions are used. During thawing, however, losses may occur. Freeze drying of fruit, which is carried out in the absence of oxygen, does not detrimentally affect vitamin C retention (Institute for Food Technologists, 1986).

Irradiation of fruits is approved in some countries. Irradiation is carried out to control maturation and insect infestation. Experimental evidence indicates that nutrient retention of irradiated fruit is comparable to that of heat-processed fruits.

Acknowledgement

The writer thanks Arrie van der Schyf of Langeberg Goods (Pty) Ltd for information on tomatoes.

References

Akhavan, I. and Wrolstad, R.E. (1980). Variation of sugars and acids during ripening of pears and in the production and storage of pear concentrate. *Journal of Food Science*, **45**, 499–501.

Bradley, G.A. (1972). Fruits and vegetables as world sources of vitamin A and C. *Hortscience*, **7**(2), 141–143.

Farrow, R.P., Kemper, K. and Chin, H.B. (1979). Natural variability in nutrient content of Californian fruits and vegetables. *Food Technology*, Febr., 52–54.

Holland, B., Welch, A.A., Unwin, I.D., Buss, D.H., Paul, A.A. and Southgate, D.A.T. (1992). In *McCance and Widdowson's The Composition of Foods*, 5th edn. The Royal Society of Chemistry, Cambridge.

Hugo, J.F. Du T. (1969). Review of literature on the health value of fruit and fruit juices. *Deciduous Fruit Grower*, **19**(3), pp. 62, 73–75.

Hulme, A.C. (1971). *Biochemistry of Fruits and Their Products*, Vol. II, Academic Press, London.

Institute of Food Technologists (1986). Effects of food processing on nutritive values, *Food Technology*, Dec., 109–116.

Jones, G.P., Briggs, D.R., Wahlqvist, M.L., Flentje, L.M. and Shiell, B.J. (1990). Dietary

fibre content of Australian foods. 3. Fruits and fruit products. *Food Australia*, **42**(3), 143–145.
Koeppen, B.H. (1974). Why we must get to know fruit sugars. *Food Industries in South Africa*. **26**(12), 23–25.
Macrae, R., Robinson, R.K. and Sadler, M.J. (1993). *Encyclopaedia of Food Science, Food Technology and Nutrition*, Vol. 3. Academic Press, Harcourt Brace Jovanovich Publishers, London, pp. 2083–2091.
Miller, J.J., Colagiuri, S. and Brand, J.C. (1986). The diabetic diet: information and implications for the food industry. *Food Technology in Australia*, **38**(4), 155–157.
Nahar, N., Rahman, S. and Mosihuzzaman, M. (1990). Analysis of carbohydrates in seven edible fruits of Bangladesh. *Journal of the Science of Food and Agriculture*, **51**, 185–192.
Potter, N.N. (1986). *Food Science*, 4th edn, Ch. 4. Van Nostrand Reinhold, New York.
Reiser, S., Powell, A.S., Yang, C. and Canary, J.J. (1987). An insulinogenic effect of oral fructose in humans during postprandial hyperglymia. *American Journal of Clinical Nutrition*, **45**, 580.
Ross, J.K., English, C. and Perlmutter, C.A. (1985). Dietary fibre of selected fruits and vegetables. *Journal of the American Dietetic Association*, **85**, 1111–1116.
Salunkhe, D.K., Bolin, H.R. and Reddy, N.R. (1991). *Storage, Processing and Nutritional Quality of Fruits and Vegetables*, Vol. I, 2nd edn, Ch. 6. CRC Press, Boca Raton, FL.
Shewfelt, R.L. (1990). Sources of variation in the nutrient content of agricultural commodities from the farm to the consumer. *Journal of Food Quality*, **13**, 37–54.
Walter, P. (1994). Vitamin requirements and vitamin enrichment of foods. *Food Chemistry*, **49**, 113–117.
Wedzicha, B.L. (1984). *Chemistry of Sulphur Dioxide in Foods*, Ch. 7. Elsevier Applied Science, New York.
Wills, R.B.H., Lim, J.S.K. and Greenfield, H. (1984). Composition of Australian foods. 22. Tomato. *Food Technology in Australia*, **36**(2), 78–80.
Wills, R.B.H., Scriven, F.M. and Greenfield, H. (1983). Nutrient composition of stone fruit (*Prunus* spp.) cultivars: apricot, cherry, nectarine, peach and plum. *Journal of the Science of Food and Agriculture*, **34**, 1383–1389.
Wills, R.B.H., Lim, J.S.K. and Greenfield, H. (1987). Composition of Australian foods. 40. Temperate fruits. *Food Technology in Australia*, **39**(11). 520–521, 530.
Wrolstad, R.E. and Shallenberger, R.S. (1981). Free sugars and sorbitol in fruits – a compilation from the literature. *Journal of the Association of Official Analytical Chemists*, **64**(1), 91–103.

3 Storage, ripening and handling of fruit
B. BEATTIE and N. WADE

Fruits are valued as nutritive foods that are pleasing to eat. The fruits discussed here are the fleshy products of floral fertilisation, which are eaten as a dessert, a salad vegetable, or as an ingredient of dishes such as stews and curries. Fruits are eaten fresh, processed into canned, frozen, and baked products, or converted into juice or jam. The raw material for processing is sometimes fruit that is unsuitable for the fresh market.

Fruits are classified by botanical or geographical relationships, similarities in fruit type or manner of cropping, or culinary use. Pome fruits (apple, nashi, pear), stone fruits (apricot, cherry, nectarine, peach, plum), and citrus fruits (grapefruit, lemon, mandarin, orange) represent botanically related species. Berry fruits (blueberry, boysenberry, loganberry, raspberry, strawberry) belong to different families but have in common a berry-type fruit that grows on a vine. Tropical fruits (banana, durian, mango, pineapple, rambutan) have a common geographical origin. Salad vegetable fruits (capsicum, cucumber, egg plant, tomato, zucchini) are grouped by culinary use.

Fresh fruits are living plant organs, a fact that is the basis of correct handling procedures. Live produce respires, exchanges water with its environment, is subject to injury by mechanical means, insects, or toxic chemicals, to disease caused by fungi and bacteria, and to metabolic disorders.

3.1 Maturity and ripening

The terms maturity and ripeness have specific meanings to the postharvest biologist. Maturity refers to the stage of development of the fruit on the parent plant. There is a minimum period of development that must be undergone by any fruit before it is ready for harvest. A mature fruit is one that either has acceptable eating quality at the time of harvest or else has the potential to ripen into a product of acceptable quality. A fruit can be mature but unripe, and indeed many fruits are harvested mature but unripe.

Bananas, kiwi fruit and avocados are examples of fruit that are picked unripe and that ripen (or are ripened) afterwards. Ripening occurs when biochemical changes convert a mature but inedible fruit into an edible product. Softening, loss of astringency, biosynthesis of aroma volatiles and conversion of starch to sugar are changes that commonly occur during ripening. The distinction between maturity and ripeness may be restated as follows. Maturation is a development process that occurs only while the fruit is attached to the parent plant. Ripening can occur on or off the plant and involves a physiological transformation of the fruit. The importance of the distinction may be illustrated by reference to the sugar content of a fruit. A fruit acquires all of its carbohydrate content from photosynthetic assimilates of the parent plant. The accumulation of carbohydrate is a developmental process, linked to maturation. The sugar content of fruits that do not undergo a rapid ripening phase cannot change appreciably after harvest. In fruits that ripen after harvest, starch, or organic acids can be converted into sugars. The amount of starch or acid available for conversion is, however, determined during maturation.

3.1.1 Climacteric behaviour

In the discussion of maturation and ripening, we have seen that some fruits undergo a phase of rapid ripening, whilst others do not. All of the fleshy fruits that are the subject of this chapter fall into two distinct physiological classes with respect to their ripening behaviour: climacteric and non-climacteric fruit.

Climacteric fruits. These fruits have a period of rapid ripening known as the climacteric. Respiration rate and heat evolution increase, sometimes dramatically, ethylene evolution often increases and the fruit softens and develops flavour and aroma. Climacteric fruits often, though not always, have stored reserves of starch and during the climacteric these reserves are hydrolysed by starch-degrading enzymes to sugars. In fruits such as mango, carboxylic acids are converted into sugars. Climacteric fruits can be induced to ripen by treatment with ethylene. Examples of climacteric fruits are given in Table 3.1.

Non-climacteric fruits. These are fruits that do not undergo a rapid ripening phase. They mature slowly whilst attached to the parent plant, and their eating quality cannot improve after harvest. Non-climacteric fruits have relatively low respiration rates that decline slowly after harvest. They produce ethylene at low rates. Application of ethylene to non-climacteric fruits has little effect other than the hastening of senescence changes, such as change in colour from green to yellow, pedicel abscission, increased susceptibility to disease and the development of off-flavours. Some non-climacteric fruits are listed in Table 3.1.

Table 3.1 Recommended conditions for storage of fruit using three temperature zones

Store at 0°C	Ripening class[a]	Store at 7°C	Ripening class	Store at 13°C	Ripening class
Apple	C	Avocado (unripe)	C	Banana (green)	C
Apricot	C			Custard apple	C
Avocado (ripe)	C	Mandarin[b]	NC	Grapefruit	NC
		Olive	NC	Lemon	NC
Berry fruit	C & NC	Orange[b]	NC	Lime	NC
Cherry	NC	Passionfruit	C	Mango	C
Fig	C	Pineapple (ripe)	NC	Pawpaw	C
Grape	NC			Persimmon (ripe)	C
Kiwi fruit	C				
Lychee	NC			Pineapple (unripe)	NC
Nectarine	C				
Nuts[b]	N/A				
Peach	C				
Pear	C				
Persimmon (unripe)	C				
Plum	C				

[a]C, climacteric; NC, non-climacteric. When climacteric fruits ripen they may evolve sufficient ethylene to intiate ripening in other climacteric fruits, or hasten senescence of non-climacteric fruits.
[b]May also be stored satisfactorily at ambient (room) temperature.

Use of climacteric class. Classification by climacteric class provides a simple guide to the general behaviour of a fruit. Banana, a climacteric fruit, may ripen during transit with resultant self-overheating. Accidental exposure to ethylene may cause such an untoward event. Oranges, being non-climacteric, cannot overheat like bananas and exposure to ethylene, although not desirable, will usually have little noticeable effect on quality unless the fruits are stored for some time.

3.1.2 Ethylene

Ethylene (ethene) is a natural plant growth regulator that is synthesised by all plants. Ethylene has many biological functions in growing plants, but in fruits it is particularly important as an agent of abscission (stem loosening), ripening and senescence. Ethylene loosens many fruits from their stems and prepares the fruit to fall from the plant or to be picked. Ethylene initiates the ripening of climacteric fruits and hastens their ultimate senescence. In non-climacteric fruits, only the senescence effects of ethylene are apparent. The effects that ethylene has on fruits are exploited in their commercial handling. Ethylene treatments are used widely to ripen or degreen fruits.

Ripening. Many fruits, of which peach and nectarine are just two examples, have begun to ripen on the tree if they are left to mature before harvest. Other fruit, such as the banana, is hard, green and unripe when mature and so can be picked in this condition and shipped to market with minimal wastage. Upon arrival at market and in response to prevailing market demand, the banana can then be induced to ripen with ethylene under controlled conditions that are optimal for good fruit quality. The technique of controlled ripening of bananas has been standard industry practice for many years and it is possible for bananas to be supplied to a customer's order so that fruit of a prescribed colour stage are delivered on a particular day. Controlled ripening is finding increasing application to other climacteric fruits that are picked unripe, such as avocados and kiwi fruit.

Controlled ripening is carried out in insulated rooms, constructed in the same way as coolrooms. Most fruits ripen best at approximately 20°C, so ripening rooms are designed to operate at around this temperature. Banana-ripening rooms are designed to run at 14–20°C. The room normally operates at high humidity, to prevent fruit shrivel, but in a 'conventional' banana room where air movement around the fruit is poor, the humidity must be lowered once ripening has begun. Ripening is initiated by adding ethylene to the room atmosphere, either by periodic manual or automatic injections or by the admission of a continuous bleed or 'trickle' of the gas. Ethylene gas is both flammable and explosive, and strict safety precautions must accompany its use. The availabililty of ethylene as a compressed gas diluted in carbon dioxide has essentially eliminated the safety hazards that have been associated with ripening rooms.

The temperature and humidity of a ripening room can be much better controlled if the 'pressure ripening' system is used. Air is compelled to enter each package or bin stacked in the room and to pass around each fruit. The technique for doing this is exactly the same as that used for 'pressure cooling', which is described in more detail below. Advantages of pressure ripening are uniform temperatures in all fruit anywhere in the room (a maximum temperature spread of less than 1°C is readily achieved) and uniform humidity.

Degreening. Although tomatoes are a climacteric fruit, they are resistant to ethylene and their preclimacteric 'green life' is only shortened by a few days by ethylene treatment. This effect is nonetheless commercially important, especially as the market requires tomatoes presented at a uniform colour stage. The differences in colouring time of a pick of mature-green to breaker tomatoes may be abolished by holding the fruit in an ethylene ripening or degreening room. Citrus fruits, which are non-climacteric, may be similarly degreened with ethylene to improve their colour for market. Degreening is an acceptable technological procedure if the fruit subjected to treatment are harvested when mature.

3.1.3 Maturity standards

A most important, but often vexing, question in the fruit industry is the use of objective criteria to decide when a crop is ready to pick. Considerable effort has been expended in devising maturity standards and standards have been set for several major crops. Difficulties remain even with the major crops and there is a lack of adequate maturity criteria for many other crops.

The application of maturity standards has been most successful with citrus and grapes, both of which are non-climacteric fruits whose composition changes little after harvest. The ratio of soluble solids to titratable acidity is widely used as a maturity criterion for these fruits. Minimum acceptable values for the ratio, accompanied by minimum acceptable soluble solids content can be set for particular varieties. The soluble solids-to-acid ratio is an excellent criterion to use with fruits where the sugars-to-acid balance is the key to acceptability and where the content of these constituents is fairly stable before and after harvest.

Maturity indices for climacteric fruits are more difficult to devise, as until the fruit has ripened only the potential quality of the fruit can be assessed. Since all the commercially important decisions about harvest, purchase and payment are made before the fruit is ripe, maturity indices should be applicable to fruit whether ripe or unripe. Soluble solids content of a climacteric fruit is dependent upon the stage of ripening. In kiwi fruit, for example, the acceptability of the ripe fruit is highly correlated with the soluble solids content of the ripe fruit, but not of the unripe fruit. When unripe kiwi fruit are tested for maturity, an estimation of the potential soluble solids content of the fruit when ripe would be a valid index. Starch content could be such a predictor, but according to research it is not. It has been suggested that the total dry matter content of the fruit at any stage of ripeness is an index of soluble solids content when ripe.

Maturity tests for fruit have been limited to a few attributes, such as soluble solids and acidity, that can be measured easily and inexpensively. The fruit is destroyed to prepare juice and the juice is assayed by refractometry for soluble solids and titration for free acidity. There is much current interest in non-destructive instrumental tests and especially of tests that could be applied to every fruit on a grading line and used as a sorting criterion. The application of such technology is in its infancy. Some current uses do exist, however, for quality rather than maturity testing. When citrus fruit are damaged by frost, their density is affected and frost-damaged fruit can be graded out automatically by an online procedure for measuring the density of each fruit. Maturity and other quality criteria could be applied by such technology, which has as its aim the packing of more consistent lines of fruit that contain only product that satisfies objective quality specifications. The application of strict quality specifications is an important part of current efforts to introduce quality management principles to the fruit industry.

3.2 Temperature and respiration

Temperature is a vital tool in the postharvest management of fruit. The metabolic rate of a fruit slows down as the pulp temperature falls and the rates of ripening and senescence are, therefore, decreased. The vapour pressure of water within the fruit tissues also decreases as the temperature falls; this reduces the rate of water loss by the fruit. Lowering the temperature also reduces infection of the fruit by microorganisms and slows the development of any existing infections. Fruit should be cooled quickly after harvest to a temperature appropriate for the particular commodity and held at this temperature subsequently, whether the product is destined for immediate retail sale, storage or processing. The importance of correct temperature management cannot be overemphasised. High temperatures are more harmful to fruit quality after harvest than before the fruit is picked. At harvest, a fruit is disconnected from its water and respiratory substrate supplies. The rate of ripening will increase in fruits where ripening has begun on the tree or vine. The green life, or time taken to begin ripening, is often shortened when fruits are harvested unripe. All of these harvest effects, which reduce shelflife, can be moderated by correct temperature management.

Many fruits keep best at a temperature that is just above the temperature at which the fruit tissues begin to freeze. The threshold temperature at which freezing begins (commonly called 'freezing point') depends particularly on the soluble solids content of the fruit, but most fruits will begin to freeze at or below $-1°C$. For practical reasons coolrooms holding such fruit are often run at $0°C$, so that the spatial and temporal variation in air temperature throughout the room can be $\pm 1°C$ without any risk of freezing. Substantial benefits in product life can be achieved by a closer approach to the actual freezing point, and it is technically possible to do this in a well-designed coolroom. Many temperate climate fruits keep best at or around $0°C$.

Other fruits, particularly those grown in tropical and sub-tropical climates, are injured if they are held below a critical temperature that is above the freezing point of the fruit. The injury so caused is known as chilling injury and fruits subject to chilling injury are called chilling-sensitive fruits, whilst fruits that are not harmed by above-freezing temperatures are said to be chilling insensitive. There is a discrete critical temperature for any particular fruit below which chilling occurs, although this fact may be obscured by varietal and climate effects. These cause variations in the critical temperature for a particular commodity, just as the freezing point varies with the soluble solids content, which, in turn, is determined by the growing conditions. The critical temperature for chilling can be as high as 13°C in the case of tropical fruits such as banana and mango, whilst other fruits, such as some varieties of avocado, may not be chilled until the temperature falls below 5–7°C. Some temperate fruits, including certain varieties of apple, may be injured by temperatures below 3–5°C and a number of stone fruits (peach, nectarine) are injured by temperatures below about 7°C, and particularly by temperatures of 2–5°C. Symptoms of chilling injury are not expressed immediately and are often not seen at low temperature. After the fruit has been exposed to a chilling temperature for sufficient time, symptoms will appear when the fruit is transferred back to ambient temperature. In general, the lower the chilling temperature and the longer the time of exposure, the worse will be the ultimate injury.

A further complication to the response of fruits to chilling temperatures is the interaction of temperature with stage of ripeness. A common symptom of chilling is the inability of a chilled fruit to ripen properly. A fruit that is already ripe when it is chilled obviously cannot manifest symptoms of abnormal ripening, and if these are the main effects of chilling on quality, the ripe fruit can be said to be more chilling resistant than the unripe fruit. Chilled pre-climacteric tomatoes do not ripen, whilst red-ripe tomatoes can tolerate moderate chilling treatment without a marked loss of eating quality. In contrast, moderate chilling of a ripe banana may also not markedly affect the eating quality of the pulp, but the fruit is usually rendered unacceptable by browning of the peel.

The distinction between chilling-sensitive and chilling-insensitive fruits is of the utmost importance in managing fruit temperature correctly. As a general rule, chilling-sensitive fruits should not be cooled below their critical temperature. In commercial handling, some compromises need to be made, and where limited storage facilities are available, it is not possible to hold each type of fruit in a mixed lot at its optimal temperature. Schemes have been devised whereby two, or preferably three, temperature regimes can be used to cater for most commodities, provided that storage for a short term only is intended. If only two storage areas are available, one can be run at 0°C and the other at 7°C. If three stores are available, the third area can be run at 13°C. Commodities can then be assigned to the two or three areas according to the dispositions suggested in Table 3.1. Particular care must be taken to clear quickly any commodities that are stored at 7°C that have a critical temperature for chilling above 7°C. If maximum postharvest life is sought, then storage must be at the optimal temperature for the particular commodity and this temperature must be maintained within a narrow tolerance. Detailed tables of optimal storage temperatures are available, and particular attention should be paid to local knowledge and recommendations that take account of local circumstances.

There are two distinct stages in the cooling of fruit, which ideally are regarded as separate operations with dedicated facilities for each. The first operation is that of cooling the fruit by removal of field heat to the temperature at which it will be shipped or stored. The concept that field-heat removal is a preliminary to refrigerated shipping or storage has resulted in the term 'pre-cooling' being used for this operation. The second refrigeration process is that of maintaining the pre-cooled product at the desired temperature by removing respiratory heat and heat that has entered the cooled space from the outside.

3.2.1 Field-heat removal by pre-cooling

Effective pre-cooling requires that large amounts of heat are removed from the product in a short time. Crops like melons, where the fruit is substantially exposed to the sun and harvest at the height of summer, may have pulp temperatures of over 50°C at harvest, so that heat of at least 2.2 kW/tonne must be removed to reach a shipping temperature of 5°C in 24 hours. Apples, which grow in part shade and are picked in the autumn/fall, may be only 20°C at harvest, so that removal of 1.0 kW/tonne will prepare them for storage at 0°C in 24 hours. In fact, the summer melon crop would be better cooled in about 12 hours, whereas apples could be allowed 2–3 days to cool. Pre-cooling must be done quickly so as to arrest deterioration of the fruit and to accommodate handling requirements like shipment. Fruit has to be pre-cooled to the carriage temperature before shipment because refrigerated vehicles and shipping containers cannot circulate sufficient cold air through tightly stacked loads to remove field heat. The refrigeration systems used during transportation are designed only to remove the heat of respiration of an already cooled product and heat that enters the refrigerated space from the outside.

The rate at which fruit can cool with adequate refrigeration is determined first by the stacking method and, ultimately, by the thermal properties of each individual piece of fruit. Fruit in a single, isolated carton will cool in several hours, especially if the lid is removed. Stacking the cartons one carton wide but several high will increase the cooling time slightly of all but the top layer, but making the stack two or more cartons wide dramatically increases cooling time. If the cartons are stacked on a pallet of about 1 m^2 in area, it will take 2 to 3 days for fruit in the centre of the stack to seven-eighths cool. The seven-eighths cooling time is the time required for fruit to cool to seven-eighths of the original difference in temperature between the fruit and the cooling medium. Seven-eighths cooling time (see Figure 3.1) may be taken as the 'practical' or approximate cooling time for fruit exposed to a coolant that is a few degrees colder than the required shipment or storage temperature. Although fruit cools faster when in bulk bins than when packed in stacked

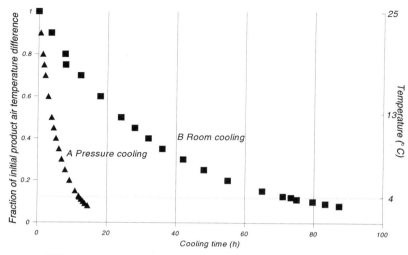

Figure 3.1 Effect of the cooling method on the way in which pulp temperature changes with time at the centre of a pallet load of fruit. The initial pulp temperature of the fruit was 25°C, and it was cooled with air at 1°C. At the start of cooling, the fractional unaccomplished temperature change (shown on the left-hand axis) was unity. When seven-eighths cooled the fraction was one-eighth or 0.125, and the pulp temperature was 4°C. Curve A is for a pallet of fruit precooled in a pressure cooler with airflow set to give a seven-eighths cooling time of 12 hours. Curve B shows the behaviour at the centre of a pallet standing in a conventional coolroom, such that the seven-eighths cooling time is 72 hours.

cartons, it still takes a day or more for fruit in the centre of a 1 m² bulk bin to seven-eighths cool. Several methods are used to overcome the delays in cooling that are caused by packing and stacking.

Pressure or forced air cooling. Heat at the centre of a stack of cartons escapes by conduction from fruit to fruit and through the layers of packaging material until it reaches the outer surface of the stack, from where it is removed by convective heat transfer, which is a much faster process than conduction. If cold air can be admitted to each carton and brought into contact with the exposed surfaces of each fruit, convective heat transfer occurs at the fruit surface, and cooling is hastened. In practice, cold air is brought into each carton or through each bulk bin by cutting ventilation slots in the ends of each carton or bin and slightly reducing the pressure on one face of the stack that is at right-angles to the flow channels formed by the ventilation slots. Cold air is then drawn into the slots on the opposite face and this air passes through the width of the stack. Convective heat exchange occurs at the interface between the surface of each fruit and the turbulent air that now surrounds that surface. A number of stacking arrangements are used to achieve pressure cooling. The simplest, which is readily adapted to an existing coolroom, is effected

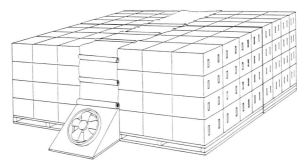

Figure 3.2 A simple system for pressure cooling pallets of cartons. The pallets are placed as shown inside a coolroom. The gap between the two parallel rows of pallets is closed off completely with a flexible cover, and a portable fan is placed at one end. The fan exhausts air from the enclosed space, so that the pressure falls. Cold room air then flows through the ventilation slots in each carton. (Adapted from Watkins, 1990.)

by building parallel stacks of palletised cartons (Figure 3.2). The cartons are ventilated at right-angles to the long axis of each stack, so that air can travel through the short dimension of each stack. An air-return passageway is created between each pair of stacks by laying a coversheet up from the floor and all the way along the gap between stacks. A small pressure-fan is fitted at one end of the air return passage, and when in use this fan exhausts air from the passage and returns it to the room as close as possible to the air collection area of the refrigeration evaporators.

Another arrangement is to construct air-collection chambers in which the fans can be mounted, either at the ends of the room for horizontal airflow, or under a false floor for vertical airflow. In theory, cooling should be faster if cold air is delivered directly through a false floor into the underside of the stacks, so that cold air flows vertically upwards. The horizontal airflow system is, however, simple and readily adapted to existing practice.

For the improvement in heat exchange between the fruit and cold air to be of any benefit, there must, of course, be sufficient refrigeration capacity available so that the added heat load that is now applied to the evaporator can be removed by the refrigerant at least as quickly as it arrives. It is not necessary to match refrigeration capacity to the initial rate at which heat is evolved by the fruit, as long as balance is achieved by about the time that the initial temperature difference between the fruit and air has fallen by half. The penalty of such a design is a small (about 15%) increase in seven-eighths cooling time, but the refrigeration plant will be only half the size that would be required to balance the two heat-exchange processes (fruit–air and air–refrigerant) at the commencement of cooling. Advantages of pressure cooling are its capital cost (which is little different to that of a well-designed conventional coolroom), flexibility (fruit can be cooled in bins or cartons, before or after packing), and the fact that the fruit is not wetted. The principal disadvantage (in some circumstances only) is that pressure cooling takes several hours.

Hydrocooling. Fruit are hydrocooled by flooding with cold water. Heat exchange at the surface of the fruit is very fast because of the high specific-heat capacity of water, and seven-eighths cooling times of 10 to 20 minutes can be achieved. The rate-limiting factor to cooling now becomes the thermal conductivity of the fruit itself, which determines the time taken for heat to travel from the centre of the fruit to the surface. Fruit with smaller diameters, like cherries, cool in 10 minutes in a hydrocooler. The particular advantage of hydrocooling in terms of managing a handling system is its speed. In terms of capital cost and energy use, however, this speed is achieved at the expense of a very large refrigeration plant, since the rate at which heat is evolved by the fruit has to be matched by the refrigeration capacity. The problems of insulating hydrocoolers often makes them less efficient than pressure coolers operating within a well-insulated coolroom.

It is almost essential that hydrocooled fruit is kept cool afterward, since the fruit, once wetted, will be very susceptible to disease unless treated with a protective chemical. The hydrocooling water itself acts as a reservoir of disease, as fungal spores washed from successive batches of fruit accumulate. The water in hydrocoolers is usually chlorinated to kill spores washed off the fruit, but chlorine treatments do not afford any residual disease control benefit, they may corrode plant, create unpleasant working conditions and cause tainting in the fruit itself.

Vacuum cooling. Precooling by vacuum uses the principle of evaporative cooling. The fruit is placed in a vacuum chamber and air is exhausted by a vacuum pump until the water in the fruit tissues starts to boil. As the water evaporates, its latent heat of vaporisation cools the fruit. Fruit can be cooled to 0°C in about 20 to 30 minutes by vacuum cooling at the expense of only a small loss in water. Although vacuum cooling is usually used to cool green leafy vegetables, which have a large surface area to volume ratio, the technology can be applied to at least some fruits. Vacuum cooling can also be used to overcome the heat-exchange problems caused by close-packing, where air or water movement around each fruit is impeded by the pack density. Vacuum cooling entails high capital investment in cooling chambers, vacuum pumps and refrigeration plant to condense the water vapour released by the process. When hydro- or vacuum-cooling is used to remove field heat, a coolroom is also required into which the precooled produce can be placed immediately and kept cool. A holding coolroom is also desirable after pressure cooling, but often the produce is held in the pressure cooling room. The temperature and humidity of a pressure cooling room fluctuate as each batch of produce is loaded and cooling begins.

3.2.2 Cool storage of fruit

Produce that is to be stored for more than a few days should be kept in a coolroom specifically designed for the purpose. The temperature of a storage room should not fluctuate and it should be closely controlled at the optimum temperature for the particular fruit. The humidity of the room should also be optimal for the product. High humidities are desirable for the storage of most fruits. Room humidity is important during storage because product water-loss during storage is directly proportional to the vapour-pressure gradient between the fruit and the room air. As the humidity of the room air approaches the internal humidity of the fruit, water-loss becomes less. It is usually not practicable or even desirable to match the internal humidity of the fruit exactly, and an average room humidity of about 90% is satisfactory for most purposes. Room humidity is not as important during precooling (by room or pressure cooling) because at this stage the fruit is hotter than the room air and the vapour-pressure gradient between the fruit and the air is primarily determined by this temperature difference. Once the fruit has cooled to the temperature of the room air, water-loss is largely controlled by the humidity difference between the fruit and the air. Although the gradient is now smaller, the longer times associated with storage come into play and harmful water-losses can occur across small humidity gradients during a typical storage period.

Several techniques may be used to obtain high room humidities. Large evaporator coil surface area and high airflow over the fins of a forced-draught evaporator enable the evaporation temperature to be kept close to the temperature of the refrigerated space, so that the condensation of water vapour on the evaporator is kept low. A useful design criterion is that the difference in temperature between air entering and air leaving the evaporator should not exceed 2°C in a 'dry coil' evaporator. 'Dry coil' forced-draught evaporators are used widely because of their many applications in refrigeration generally. An alternative system that has particular application to fruit storage uses water as a secondary refrigerant. A fan circulates room air through a fine mist of chilled water so that the air is cooled to the temperature of the water and simultaneously saturated with water vapour at that temperature. A particular advantage of this system is that the air can be saturated at any operating temperature. Both the dry coil and the water mist systems can be designed so that air leaves the heat-exchange surface saturated with water vapour, but as the air gains heat (from the product and from the outside environment) the increase in air temperature will result in a decrease in relative humidity. It is simply not possible to transfer heat and keep the humidity at saturation, even if it was desirable to do so.

Another important aspect of room design that affects humidity is insulation. Heat that leaks into a coolroom from the outside raises the temperature of the air and lowers the relative humidity. This effect is minimised in a well-insulated room, which also requires less electricity to run. The walls, ceiling and floor of a storage room should be well insulated. Various standards for permissible heat penetration have been suggested. A transfer rate of 10 W/m^2 should probably be considered the absolute maximum that can be tolerated.

Another method of obtaining a high humidity for storage is the 'jacketed room' system. This method combines the principles that have been outlined above. An uninsulated cubicle is built within the main coolroom and produce is cooled in this cubicle. The penetration of external heat into the cubicle is kept to a minimum because the ceiling and walls are 'jacketed' with an envelope of cold room air. The nature of the refrigerant-to-air heat exchange is made irrelevant because the room air is now serving as a secondary refrigerant. The air inside the cubicle transfers the respiratory heat of the fruit and a very small amount of transmitted heat to the room air, with the exchange taking place on the exposed surfaces of the cubicle. Because this surface area is very large, the heat can be exchanged across a very small temperature gradient and so condensation onto the exchange surface is minimised and the humidity in the cubicle is kept high.

All of the methods for obtaining high humidity that have been described need to be incorporated into the original design. Where an existing room is running at too low a humidity, less satisfactory remedial measures can be taken. The most usual remedy is to install a spray humidifier that adds a mist of water to the room air. Humidifiers may cause wetness problems in the room and the evaporator coils in a room operating below 2°C will ice up faster.

The maintenance of a uniform temperature and humidity throughout a storage room requires that there be good air circulation. The produce should be stacked with at least 100 mm clearance from the walls and floor and with gaps between adjacent pallets or bins so that air can move freely. The room air should be circulated continuously by fans. Less fan power is needed than during room pre-cooling and it is customary to switch off some of the forced-draught evaporator fans during storage, whilst leaving the rest to run continuously (rather than cycling with the room thermostat). In water-shower heat-exchange systems, the fan must run continuously anyway. An air circulation rate of 15 to 30 room volumes per hour is generally sufficient during storage.

3.2.3 General requirements of a coolroom

The full particulars of coolroom design are far too detailed to describe here, but some important general principles should be mentioned. Materials and methods of construction will not be dealt with since these vary considerably according to local requirements and procedures. The structure must be well insulated and sealed against the ingress of outside air. Doors must close tightly, and if they will be open for long periods during loading and unloading the door entrances should be protected by plastic flaps, streamers, or air-curtains to reduce the ingress of outside air. There must be sufficient working space within the room for machinery, such as forklift trucks, to operate freely. The floor must be strong enough to carry the mass of product to be loaded and the live load of vehicles. A most important element of the structure is the vapour barrier. This is a barrier against the permeation of water vapour into the wall insulation. In the absence of an efficient vapour barrier, water vapour will migrate into the wall on the hottest side and condense adjacent to the cool side. The condensate will reduce the efficiency of the insulating material and cause deterioration of the structural elements of the wall. Such deterioration can be quite rapid and very costly. The correct location of the vapour barrier depends upon local climatic conditions. In warm climates, the outside temperature will almost always be higher than the temperature within the room and so the barrier should be placed on the outer side of the insulation. If a room operating in a cold climate will usually be warmer inside than out, then the barrier should be placed on the inner side of the insulation. The vapour barrier is made out of material that is impermeable to water vapour, such as steel, plastic film or bitumen-treated papers. Where coolrooms are built from panels made by laminating plastic foam insulation between two sheets of steel, one of the steel sheets will serve as the vapour barrier (being the outer or inner sheet, according to climate). All joints between panels must, of course, be well sealed with a durable and waterproof sealant. Once a secure vapour barrier has been constructed, it is important that the barrier is kept intact. Any penetrations of the barrier should be avoided or, if necessary, the penetration must be well sealed with the sealant.

3.2.4 Mixed storage

The need to carry or store different commodities in the one place is a common problem. In evaluating the compatibility of different products, consideration must be given to their respective temperature requirements, relative sensitivity to ethylene and the adsorption of undesirable odours. Detailed information on product compatibilities is available. The question of temperature compatibility has been discussed already and suitable courses of action outlined. The interactions between high and low ethylene-producing commodities are more complex, but the ethylene produced by apples will, for example, cause green bananas to start ripening, kiwi fruit to soften, cucumbers to turn yellow and carrots to turn bitter. Produce such as onions that have strong aromas are likely to taint other produce, packaging materials and storage rooms. Potatoes are apt to cause an earthy taint and durians a complex taint.

3.3 Storage atmospheres

The storage life of a fruit can be affected quite substantially by the composition of the atmosphere in which the fruit is stored or carried. The storage atmosphere can be either beneficial or detrimental, and practical use is made of the beneficial effects. The harmful effects arise if the oxygen concentration approaches zero, as can happen at the centre of a stack of hot fruit, or if carbon dioxide or ethylene accumulate because of poor ventilation. The storage lives of quite a number of fruits can be increased by reducing the O_2 concentration, often to about 2–3%. A smaller number of fruits keep better if the CO_2 concentration is increased, usually to just a small percentage, but sometimes up to about 20%. Most commodities that respond favourably to increased CO_2 will respond best if O_2 is simultaneously lowered. In some cases, such as peach and nectarine fruits, there is little or no benefit from lowering O_2, but substantial benefit from relatively high concentrations of CO_2. It is most important to stress that too low a concentration of O_2 or too high a concentration of CO_2 will injure fruit. The threshold concentrations for injury vary with commodity and temperature. Fruits are a bulky mass of respiring cells. The internal atmosphere of a fruit is always different from that of air because respiration prevents the internal atmosphere from equilibrating with the outside atmosphere (air). Changes in the O_2 and CO_2 concentrations of the outside air make the internal atmosphere of the fruit even more extreme. If a fruit is placed in an atmosphere containing only 3% O_2 the internal atmosphere of the fruit will contain only a fraction of a per cent of O_2 at most. This small amount of O_2 will usually be sufficient to sustain normal aerobic respiration. If the outside O_2 concentration falls below about 1%, it is likely that the resultant trace of O_2 left in the fruit will be unable to sustain aerobic respiration and anaerobic respiration will take over. Anaerobic respiration cannot completely oxidise sugars, and products such as ethanol and acetaldehyde accumulate. The accumulation of potentially toxic metabolites, the reduced yield of energy, or both, lead to cell injury and tissue death. Too high a concentration of CO_2 is also harmful. As little as 1% CO_2 in the outside atmosphere may harm some varieties of citrus fruit, apple and pear, whereas other apple varieties will benefit from 5%

CO_2 or even more and strawberries and a number of stone fruit will tolerate 20% CO_2. Carbon dioxide has many effects on cell metabolism, including inhibition of the respiratory pathway such that symptoms of injury similar to those caused by lack of O_2 are observed.

The beneficial effects of low O_2 and elevated CO_2 are varied, but where present they confer longer storage life and maintain quality. Apples can be kept crisp, juicy and green for up to twice as long as in conventional cool storage. Kiwi fruit and bananas can be kept hard, green and unripe. Peaches and nectarines can be kept firm, juicy and free of mealy breakdown. Mangoes keep longer than otherwise in such altered atmospheres. Other fruits, particularly those of the non-climacteric category, show less or no response to changes in the storage atmosphere. Citrus fruit, cherries and strawberries, for example, do not respond dramatically to atmosphere changes. The beneficial effect that 20% CO_2 has on cherries and strawberries is probably caused by suppression of fungal growth.

The means by which changes in atmosphere composition improve storage life are not completely understood, but an important element of the response is control of the internal ethylene system. Molecular O_2 is required by the enzyme that produces ethylene from its amino acid substrate, and the lower O_2 concentrations that give beneficial storage responses inhibit ethylene synthesis by the fruit. Carbon dioxide competes with ethylene for its binding sites in the fruit cells and the small percentage of CO_2 that has beneficial effects is often sufficient to greatly reduce ethylene binding. The concentrations of O_2 and CO_2 that improve storage life also reduce fruit respiration rates, in common with cool storage, and it is true that for a particular type of commodity reduced respiration rate is often associated with increased storage life.

Fruit may be stored under either a controlled or modified atmosphere. A controlled atmosphere is maintained within close tolerances by analysis and adjustment as required. Control may be manual or automatic. Modified atmospheres are not under close control and they are usually generated by the respiratory activity of the fruit itself.

3.3.1 Controlled atmosphere (CA) technology

The first requirement is for a gas-tight storage room. It is difficult to build such a room and the methods used vary widely. As with coolrooms, local building practices and codes determine the methods used. The principle of a CA store is that the inner faces of the room should be lined with material that is impermeable to gases. The requirements for such a gas barrier are similar to those of a vapour barrier, and local climate determines whether the gas and vapour barriers are separate or combined. In cold climates where the vapour barrier is placed inside the insulation, one barrier will serve both purposes. In warmer climates, the vapour barrier is still placed outside the insulation, while the gas barrier is best placed inside. Materials like steel and plastics make good gas barriers, although coatings of various types are also used. Joins between sheet claddings are sealed with flexible sealants and covered with mouldings. A technique for applying a plastic gas barrier is to spray the interior of the store with polyurethane foam. A sufficient thickness of foam forms a barrier to gas permeation and also serves as thermal insulation. Special doors with gas-tight seals are required. A common problem with such rooms is that movement of the structure opens leaks in the gas barrier. Structural movement caused by pressure differences between the inside and outside atmospheres must be prevented by allowing the pressures to equilibrate. A water-sealed pressure relief pipe should be inserted through the wall of all CA stores. Movement caused by thermal expansion and contraction and by other factors, such as wind, is not easily accommodated.

3.3.2 Atmosphere generation

Some common ways of generating the controlled atmosphere are listed below.

Using fruit respiration. Sealing the packed store allows fruit respiration to modify the atmosphere, which can then be controlled within the desired limits. It is only practicable in a refrigerated store to let the fruit generate the atmosphere once, immediately after the store is loaded, when the fruit is still warm and respiring rapidly. Cold fruit will not be able to restore an atmosphere that is lost through leakage or the need periodically to open the room and remove stock.

An open flame hydrocarbon fuel burner. Air is burnt in a gas flame and the combustion products (unburnt O_2, CO_2 and moisture) are discharged into the room. If fruit respiration causes O_2 to fall, outside air is bled in. If CO_2 increases, the room air is recirculated through a CO_2 scrubber. If leakage causes O_2 to increase, more outside air is burnt and used to flush the room. Open flame burners are only economical with small rooms because of their inability to burn O_2 that leaks into the room.

Catalytic O_2 burner. These generators overcome the inability of open flame units to use the O_2 in recirculated room air. They contain a bed of heated catalyst that allows fuel to burn in the low O_2 concentration of the room atmosphere.

Air separators. These take compressed air and fractionate the constituent gases by their molecular sizes. In one type of separator, air flows into hollow fibres made from a membrane in which O_2 permeates faster than

N_2. By the time the air reaches the opposite end of the hollow fibre, much of the O_2 has diffused out, leaving a N_2 rich gas. In the pressure swing adsorption separator, compressed air enters the bottom of one of two packed beds of adsorbent. The bed adsorbs O_2 so that a N_2 rich gas is discharged at the top of the bed. When the adsorptive capacity of the first bed nears saturation, the separation is switched to the second bed while the first is regenerated.

Liquid N_2. The O_2 concentration in a room can be reduced very quickly by flushing with N_2 that is injected into the room as liquid. Liquid N_2 is usually used only to establish the initial atmosphere and not for subsequent maintenance of low O_2.

Scrubbers. The use of scrubbers to control CO_2 in the atmosphere has been mentioned. Many different types of scrubber are used to remove CO_2. The majority rely on the reaction between CO_2 and alkalis, and one type relies on the solubility of CO_2 in water. The air separators that have been described as generators also function as CO_2 scrubbers, since CO_2 partitions in a manner similar to O_2.

Safety hazards. Hydrocarbon burners are a potential safety hazard and they should be designed with safety devices and used by properly trained and experienced operators. It is possible for a storage room to be filled with flammable gas as a result of burner malfunction. Air separators pose no such safety risks and by running them in a flow-through mode, so that the store atmosphere is continuously flushed, it is possible that fruit quality may be improved by the removal of harmful fruit volatiles.

Analysis of gases. The store atmosphere may be analysed manually by the use of apparatus such as the Orsat volumetric gas analyser, or with instruments such as the infrared analyser for CO_2 and a paramagnetic or fuel cell analyser for O_2. Any of the electronic instruments can be set up to sample and read the room atmospheres regularly, and record the results. The analyser outputs may also be used to operate generators and scrubbers, so that the atmosphere is controlled automatically. Regardless of the technology used, accurate records should be kept of store atmospheres and store and fruit temperatures, so as to assist in the diagnosis of any quality problems that may arise in fruit after removal from storage.

Handling of fruit. The selection and handling of fruit for CA storage is of the greatest importance in determining ultimate fruit quality. Unless fruits are mature when picked, they will never ripen into a product of acceptable quality. There are no exceptions to this rule, which applies whether the fruit goes to the domestic market or is stored. The stage of ripeness at which fruit enters storage is, however, critical. The efficacy of CA storage is reduced when fruits have entered the ethylene climacteric and have begun to ripen before storage. Fruits that are preclimacteric but mature will respond best to CA storage. It is common for harvest to be delayed until fruit size and colour appear satisfactory. The fruit may well have begun to ripen before these attributes are judged as satisfactory, with the result that the harvest will have a reduced storage life. Disorders will develop and quality will deteriorate if storage is continued for a longer time.

Speed of cooling. The speed with which fruit is cooled and CA conditions are established is also critical if good results are to be obtained. Postharvest delays in attaining storage conditions allow the fruit to begin ripening, if it has not already done so, and for ripening changes to develop. Ripening usually begins sooner and develops faster once the fruit has been detached from the parent plant. The aim of good storage practice is, therefore, to pick the fruit and place it in an optimal storage environment as soon after as possible. With fruit like bananas, storage conditions should be established within a day of harvest, while with apples, the fruit should be cooled and under CA within a week of harvest. It is preferable to delay harvest if need be to achieve these time requirements, because ripening of attached fruit is slower. Too long a delay will result, however, in the fruit becoming too ripe for any useful storage.

Segregation of fruit. Segregation of fruit within the store is another management tool. It is possible with at least some fruits, like apples, to classify each line of fruit coming into the packing house as suitable for 'short', 'medium' or 'long-term' storage. With apples, maturity and ripeness criteria like pressure test, soluble solids and iodine starch test may be used to assign lines of fruit to a storage class. Calcium analyses have also been used to predict the probability of storage disorders in apples. The store is then emptied according to the anticipated storage life of the several storage classes.

Evaluation of shelflife. Regular monitoring of the stored fruit by withdrawal of samples and evaluation of their shelflife is another useful tool. A small airlock can be installed in the room so that samples of fruit can be removed from outside without breaking the atmosphere. The samples are allowed to ripen at room temperature and assessed for the incidence of disorders and other quality defects. At the first sign that quality is starting to deteriorate, the room should be opened and the affected fruit should be sent to market.

3.3.3 Modified atmosphere technology

Modified atmospheres have particular application as an in-package treatment for use during transport, and sometimes in a storage room. The atmosphere is generated by lining each package with a material, such as plastic film, that is only partially permeable to gases. The fruit is packed in a carton lined with a bag made of the plastic film, and the top of the bag is either folded down or tied off. The plastic film restricts the outward movement of CO_2 and water vapour evolved by the fruit, and the inward movement of O_2. Provided that sufficient O_2 can cross the film barrier to support aerobic respiration, the atmosphere within the bag will acquire at constant temperature a steady-state composition such that $(21 - V_o)/V_c = (P_c/P_o)/R$, where V_o and V_c are the concentrations of O_2 and CO_2 respectively, as per cent by volume, P_c/P_o is the ratio of the permeability coefficients of CO_2 and O_2, respectively, and R is the ratio of CO_2 evolved by the fruit to O_2 consumed. The steady-state composition of the atmosphere inside a plastic film package is, therefore, a function of the respiratory gas ratio (a biological property of the fruit) and the selectivity factor of the film for CO_2 relative to O_2 (an attribute of the film). The selectivity ratio (P_c/P_o) of a hole is close to unity, whereas the ratio for polymeric films is invariably greater than unity. The selectivity ratio P_c/P_o for low-density polyethylene, for example, is commonly about 5. Suppose that the desired storage atmosphere contains 5% O_2 and less than 10% CO_2. In a package where $P_c/P_o = 1$ and $R = 1$, the resultant atmosphere will contain 5% O_2 and 16% CO_2. If $P_c/P_o = 3$, the atmosphere will be 5% O_2 and 5.3% CO_2. This concept is shown diagrammatically in Figure 3.3. In a bag closed by folding, or made of a film that has been manufactured with micropores or small punctures, the selectivity ratio will be close to unity. If the bag is tightly closed and gas transfer is by permeation through the film, the selectivity ratio will approach that of the film, and the CO_2 concentration will be reduced compared with a similar package with a selectivity ratio of unity. The selectivity ratio of a package rarely reaches that of the film from which it is made, because of leaks at the bag closure and punctures in the bag.

A major limitation to the use of modified atmospheres is the effect of temperature on the atmosphere. As temperature increases, so does the respiration rate of the fruit and the permeability of the film to gases, but the temperature coefficients of both processes are different, so that a more severe atmosphere invariably develops if temperature rises. Fruit that has been shipped under refrigeration in modified atmosphere packages is liable to injury if the consignment is allowed to warm up at its destination.

In the preceding discussion, controlled atmospheres have been treated as a technology appropriate to storage rooms, and modified atmospheres as having particular application to transport. This presentation reflects

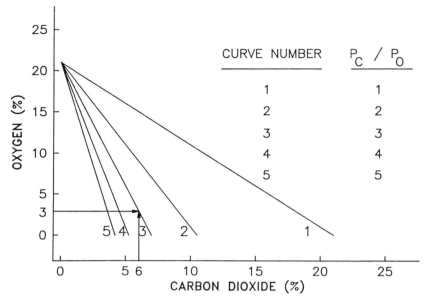

Figure 3.3 Principle of modified atmosphere packages that rely on film permeability. The graph refers to an ideal package without holes. Each line defines the combinations of package O_2 and CO_2 concentrations that can be attained with a particular film selectivity ratio (P_c/P_o), if the ratio R of CO_2 evolved by the fruit to O_2 consumed is unity. If, for example, $P_c/P_o = 3$, one of the possible atmospheres that may be attained is 3% O_2 and 6% CO_2. The actual atmosphere attained will depend on the mass and absolute respiration rate of the fruit, and the surface area of the package. (Adapted from Wade and Graham, 1987.)

common commercial practice, but alternative applications do exist. Atmospheres in large storerooms have been generated by respiration, and controlled by gas permeation units that use polymeric films. Similarly, controlled atmosphere equipment has been attached to shipping containers and used to control the container atmosphere during transit.

3.4 Maintaining quality

The fulfilment of consumer demands for fresh high-quality fruit and other horticultural produce has only been possible because of the use of the technology described here. Despite this, losses in produce still occur. The value of produce to the producer, wholesaler, retailer or processor may be lowered or completely lost as a result of postharvest mismanagement. The intangible losses to the community represented by consumer dissatisfaction with poor quality and highly priced produce are probably greater.

There are three major causes of loss during the storage and handling of fruit. These are:

- disease caused by pathogens: fungi and bacteria
- non-pathogenic disorders caused by disturbances in the normal metabolism of the fruit; some of these disorders may be caused by adverse environmental conditions such as extremes of temperature
- injuries caused by mechanical force, insects or toxic chemicals.

Diseases caused by fungi have an incubation period between the time of infection of the produce and the appearance of symptoms. This delay varies from a few hours to a number of weeks and is affected by the speed at which the pathogen can grow, by the susceptibility of the fruit to the pathogen, and by temperature, atmosphere and humidity.

The term disorder is usually applied to those problems in fruit not caused by pathogens. Symptoms are caused by fruit reacting to some kind of stress connected with temperature, humidity, atmosphere and time from harvest. In many cases, disorders cause loss of quality in the fruit rather than the complete loss that rots can cause.

Physical injuries to fruit tissue can be caused by a number of factors such as mechanical damage (hail, dropping, squashing, rubbing, tearing, puncturing), exposure to extreme high or low temperatures or by insect damage.

The principles that lie behind good postharvest management take into consideration the protection of fruit as a living organism. Losses and deterioration in quality can be greatly reduced by attention to field, packing house, storage and retail management.

3.4.1 Disease

Disease is a major cause of loss in the harvested fruit crop. Most postharvest disease is the result of infection after harvest, but sometimes infection occurs in the field. The infection (usually fungal spores) may be acquired in the orchard or after harvest, through contact with contaminated picking receptacles, boxes, bins, wash water, sorting equipment and the like. The most important single event that causes postharvest disease is mechanical injury, which allows invasion by wound-infecting organisms.

All pathogenic diseases result from an interation between a parasite, the environment and a susceptible host. Most fruit parasites are fungi. The source of fungal inoculum is usually spores. Spores may be conveyed in the orchard by wind, rain-splash, physical contact or insects, and after harvest by the containers etc. Important elements of the postharvest environment are humidity, free water and temperature. Most fungal spores require high humidity or free water to germinate. Fruit kept in a confined space (such as a carton) are surrounded by a humid microclimate, and spores resident at the site of an injury have the additional assistance of leaking cell sap. The rates of both spore germination and infection increase with increasing temperature up to a maximum, after which the rates decline, usually quite rapidly. Host susceptibility depends in the first instance on genotype. Most fruits are resistant to most pathogens, and a high degree of host specificity occurs. For example, *Penicillium* species are important causes of postharvest loss, but pathogenicity is quite specific. *P. expansum* is a pathogen of apples and grapes, but this species will not infect citrus. *P. digitatum* and *P. italicum* both infect citrus, but not grapes or apples. Susceptibility depends secondarily on the physiological state of the fruit, which varies according to stage of ripeness and handling history. Ripe fruits are more susceptible to infection than unripe fruits. Fruits that have been chilled are more susceptible to infection, and sometimes diseases are the most obvious symptom of chilling. Mechanical damage during handling provides wounds through which infection can occur. Latent infections occur when fungal spores form resting stages on the fruit surface in the field. Infection hyphae may penetrate the fruit surface, but no further development of the infection occurs until after the fruit has started to ripen. The anthracnose diseases of tropical fruits are examples of latent fruit diseases. Latent infections may occur in flowers before the fruit has developed. The grey mould rot (*Botrytis cinerea*) that affects many fruits can arise from latent floral infections.

Postharvest disease causes wastage of fruit, but postharvest pathogens that have not infected the fruit or have only just initiated infection can cause wastage of fruit products. Fruit pathogens often secrete pectolytic enzymes that are used to secure entry to the host. Some of these enzymes are heat resistant, and if fruit is canned that carries a surface microflora of pectolytic enzyme-secreting fungi, enzyme activity left after thermal processing will macerate the product.

Control of postharvest fruit disease is achieved by integrated good practice both before and after harvest. Field disease-control measures reduce the carryover of disease and inoculum into the packing house. Cultural practices that let air and sun into the trees and reduce warm, moist microclimates are helpful. Plant nutrition can affect disease. Excessive applications of nitrogen can increase pre- and postharvest disease, probably by lowering fruit calcium content. Fast cooling after harvest arrests infections that would otherwise form at wound sites and in ripening tissues. Good handling that minimises mechanical damage reduces opportunities for infection. Pickers should know the correct method of handling fruit. They should have short fingernails and preferably wear gloves. Fruit should be picked with a swift movement that does not take wood or leaf. Picking bags should not be overloaded nor the fruit dropped when delivered to field lugs. The packing house and coolrooms should be kept clean. Floors should be swept regularly and at the end of each day's packing. Dirt and dust should be kept out of the building, and spilled or rejected fruit should be gathered and placed in a covered container that is replaced daily. Neglected mouldy fruit can discharge clouds of spores. The equipment used to dump and sort fruit should be well maintained and free from defects that might injure fruit. Steam clean equipment each season. All belts and conveyors should be swabbed down each day with a sanitising solution. Chlorine bleach preparations or alcohols can be used for sanitation, provided that care is taken against the corrosive properties of chlorine. Sometimes the interior of a coolroom has to be washed with a sanitiser because of mould growth. The primary cause of such problems (which are often associated with excessive condensation) should be investigated and remedied.

Fungicides can be applied to fruits after harvest in a dip before sorting, as a spray drench on the line, or as an additive to wax treatments. Fungicides may be progressively inactivated in dips or recirculating sprays by binding to soil or sap. Some formulations are unsuitable for application through spray nozzles, and mixing fungicides with wax reduces efficacy. Postharvest fungicides usually have to prevent infection, or resumption of infection, by an inoculum already resident on or in the fruit. The length of time between harvest and chemical treatment has a critical effect on efficacy of disease control, since infection is hastened by harvest wounds and ripening. Most fungicides should be applied within 24 hours of harvest when fruit are at ambient temperature (Figure 3.4). A longer delay may be permissible if the fruit is cooled quickly, since infection is slowed at low temperature. Treatment efficacy depends in part on uptake of fungicide at wound sites, where it prevents infection, often by interfering with spore germination or the growth of newly formed hyphae. Some postharvest fungicides, such as the benzimidazole group and the imidazole ergosterol biosynthesis inhibitors, have an eradicant action whereby existing infections are controlled or suppressed. These chemicals can penetrate waxy cuticles and reach seats of infection within the fruit. Each postharvest fungicide has a specific spectrum of activity and is only suitable for the control of particular diseases. Postharvest fungicides are food additives that must be used in accordance with the applicable legislation.

Fungi frequently develop resistance to postharvest fungicides. Populations of resistant strains can build up quite quickly, so that the fungicide in question suddenly loses efficacy in a particular packing house. The presence of resistance is diagnosed by laboratory testing of isolates of the suspected resistant strain. Resistance-avoidance strategies rely on the use in rotation of several fungicides in both the field and packing house, and strict adherence to sanitation measures in the packing house. Waste fruit that go mouldy or rot in the vicinity of the packing house may be multiplying a resistant strain that will then infect fruit entering the

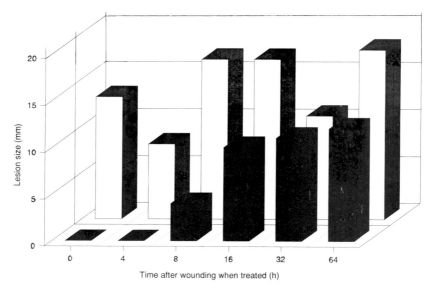

Figure 3.4 Effect on disease control of delays in application of a fungicide after fungal spores have been introduced at wound sites. Muskmelons were wound-inoculated with spores of the fungus *Geotrichum candidum*. Random samples of the fruit were dipped in either water (control) or a solution of the fungicide guazatine at subsequent intervals. The average size of sour rot lesions was measured 5 days later. The diagram shows that the efficacy of disease control was reduced when the delay between wounding and treatment exceeded about 8 hours. (Adapted from Wade and Morris, 1982.) □; water. ■; guazatine.

premises. It is best to prevent resistance developing in the first place, since the sudden appearance of resistance will cause loss, and withdrawal of the affected fungicide provides no assurance that the resistant strain will disappear.

Several other disease-control strategies are currently being investigated to reduce fungicide use. Heat treatment can be effective, but it is often difficult to achieve satisfactory control without injuring the fruit. Biological control has been achieved by dipping fruit in suspensions of bacteria that grow on the fruit surface and inhibit fungal pathogens. Such a treatment has yet to be approved by health authorities. Strategies that promote wound healing can reduce disease caused by wound-infecting pathogens.

3.4.2 Disorders

Fruit may be affected by breakdown or blemish that is caused by an abnormality in the physiology of the fruit itself, rather than by the primary intervention of a pathogen. Such an injury is called a disorder. Disorders may have an external cause, but unlike diseases this cause is environmental rather than pathological. Although it is conventional to divide fruit breakdown into disease and disorder, abnormal fruit physiology is common to both conditions, and the distinction lies really with the predisposing cause, which may be pathological, environmental or wholly internal. Some examples of fruit disorders are discussed below.

High-temperature/solar injury disorders. Most fruits develop sunburn if overexposed to solar radiation. Vein tract browning of melons is caused by preharvest solar ultraviolet radiation, followed by postharvest desiccation. The sun and other sources of heat may also cause thermal injury, manifested as abnormal softening and an inability to ripen normally.

Low-temperature injury. All fruits are liable to injury if their tissues begin to freeze. The resultant tissue injury becomes visible when the frozen tissue thaws, and the extent of injury depends on the rate of thawing. The phenomenon of chilling injury, which was discussed in the section on temperature, is another important disorder. The injuries that occur in tropical fruits stored below about 10–13°C are the chilling injury disorders. Storage disorders that occur in temperate fruits like apples and peaches below about 5°C have not yet been shown to arise from the same primary chilling event as the tropical fruit disorders, and it is not even clear if temperature causes the injuries observed. It may well prove, however, that temperate fruits are also subject to chilling injury, though at lower threshold temperatures than tropical fruit. Breakdown, superficial scald and soft scald of apples, and cool storage or mealy breakdown of peach and nectarine fruits are examples of storage disorders of temperate fruits.

Nutrient deficiency or imbalance. Fruits are peculiarly susceptible to disorders caused by low calcium or by a nutrient imbalance involving calcium. Calcium is carried into fruits and leaves by the transpiration (water) stream, and fruits, unlike leaves, have low transpiration rates. A classic calcium deficiency disorder is bitter pit of apples. In extreme cases corky 'pits' develop in affected fruit on the tree, but symptoms usually appear after postharvest storage.

Control measures for disorders are suggested by the cause, where this is known. High-temperature injury is avoided by maximising leaf cover (by pruning and disease control) and, in some cases, by erecting covers over the crop and even whitewashing fruit on the vine. Picked fruit are susceptible to heat injury and should be kept shaded. Postharvest low-temperature injury is generally avoided by holding fruit above the critical threshold temperature for injury. The loss in shelf life that would otherwise occur at this higher temperature might be prevented by another technique, such as controlled atmosphere storage. The symptoms of chilling injury can sometimes be alleviated. Some examples of alleviation are the dipping of citrus fruits in a hot suspension of benzimidazole fungicide before exposure to cold, and the waxing of fruits before refrigeration. Apples are dipped in antioxidants to control the storage disorder superficial scald. Calcium-related disorders respond poorly to soil amendment with calcium, because little of this calcium can reach the fruit. Spraying fruit in the orchard can be effective, but coverage must be complete. Most success has been achieved with postharvest application of calcium salts by dipping.

3.4.3 Injury

Mechanical injury. Mechanical injury is an important cause of down-grading and wastage of fruit. Damage may occur before harvest from wind rub, hail, pressure by adjacent fruit as they expand and high turgor. Most damage occurs during and after harvest, from manual handling by pickers and packers, mechanical handling on dumping and sorting machinery, or during transport. Wastage may be caused by physical damage or become the ultimate consequence if injury is followed by infection.

Impact bruise injury occurs by fruit-to-fruit or fruit-to-hard surface contact when fruit are dropped or bounced. Falls of only several millimetres can be sufficient to cause bruising. Compression bruising is caused by a static load, such as the pressure of overlying fruit in a stack, or the pressure applied by the walls or lids of packages or bins. Bruise injury takes at least several hours to appear, and it may take weeks. Bruising is sometimes only apparent when a fruit is cut. Other types of mechanical injury include punctures (from fruit stalks, fingernails or poorly maintained field boxes, bins and handling equipment), abrasion injury, which occurs when fruit rub together when moving on a conveyor or during transport, and cracking, which occurs when the fruit skin fails under excessive internal turgor pressure. The susceptibility of fruit to mechanical injury depends on several factors. Fruit pulp temperature affects susceptibility to bruising. The relation is often complex, and impact and compression bruising may each respond quite differently to temperature. Handling protocols may specify temperature limits for operations like sorting, so as to minimise bruise injury. Some fruits should not be picked or handled when their turgor pressure is high, such as in the cooler times of day, when wet with dew or condensation or after rain. Fruit may be too soft to handle. The decision of whether to grade fruit before or after storage will often depend on how soft it will become during storage.

Insect injury. Insects infest fruit and cause damage in the field. Growth malformations, russetting, scars and other blemishes are common symptoms. Some insects lay eggs in fruit, and when hatched the larvae consume the fleshy tissue. The best place to control insects is in the field, so that fruit is protected as it develops. Control is by integrated pest management, where biological and chemical measures are used to best effect. Postharvest disinfestation treatments are applied to fruits so as to satisfy quarantine requirements. The purpose of quarantine is to prevent the introduction of harmful pests, and high levels of security are required. Fumigation schedules based on halogenated hydrocarbons have been used as quarantine treatments for many years, but these fumigants are being withdrawn from use. Other treatments that are used include cold disinfestation, where the fruit is coolstored for about two weeks, and heat disinfestation, where it is heated to about 50°C, often for an hour or more. Gamma radiation is effective but unwelcome to the consumer. Alternative fumigants and controlled atmospheres are other possibilities that are currently under study.

Packaging and mechanical handling. Correct packaging and handling procedures prevent mechanical damage to fruit during transport and distribution. Fruit can be handled in bulk without packaging if a small amount of damage is acceptable. Bulk handling is appropriate where fruit comes straight from the field to a processing plant, but even fresh market fruit can be shipped in bulk if it is resorted before sale. The technology used to pack and handle depends upon the cost and availability of materials and labour, the presence or absence of a crate-exchange system for reusable crates, and the means available to dispose of single-use packages. It is preferable that single-use packages be made of materials that can be converted for recycling. Moulded liners may be used within cartons to reduce fruit-to-fruit contact and movement. Packages (cartons or crates) are best handled as unit loads on pallets or slip sheets. Various gluing or strapping systems are used to hold stacks of cartons together, so that they can be moved as a single unit on hand- or fork-life trucks. Unit loads are much less vulnerable to mechanical damage than individual packages. Unit loads cool slowly, however, unless they are ventilated for pressure cooling. The usefulness of unit loads is also diminished unless there is an integrated transport and materials handling infrastructure designed for them. It is inefficient to have to break a unit load apart because vehicles will not accommodate them or because equipment to load or unload is unavailable.

Managing the postharvest system. A number of principles and technologies used in ripening, storage and handling fruit for the market have been discussed in this chapter. It must be remembered, however, that the reason why these processes are used is to provide the end market with fruit the market wants. For the product to come out of the packing house with value, it has to go through a chain of these different processes. Many of these processes or technologies are sophisticated and require a thorough knowledge of the operation by managers and by packing house staff. The handling of produce after harvest should be looked at as a system, where each component such as storage, packing and fungicide treatment contributes to the quality and subsequent value of the final product. If the final product is not what the consumer wants, the value is decreased to all participants. A major limitation in the success of marketing fruit is the poor management of the technology of the postharvest system.

Much effort is now being directed in the horticultural industry to the introduction of quality management principles into managing the postharvest system. Quality management uses these systems analysis techniques to establish a planned and systematic control over all of the activities in the packing house that influence product quality. A quality management system in a packing house focuses on the prevention and correction of defects at the earliest possible stage in the production process. The system calls for a continual sampling of the product at every stage during its handling to see if it conforms with the reference standard at that particular stage. At the same time, the equipment or technology being used is also monitored against a standard to ensure that the technology is doing what it is supposed to do. For instance, the quality management system would ensure that if fruit of a particular measurable maturity was required, then only that fruit would be used. A quality management system would also ensure that a controlled atmosphere room would only run at the specified gas and humidity concentration.

Quality management is the major method by which the postharvest handling of fruit using the technology described can be utilised most efficiently and profitably.

Further reading

Beattie, B.B. and Revelant, L.J. (ed.) (1992). *Guide to Quality Management in the Citrus Industry*. Australian Horticultural Corporation, Sydney, Australia.

Beattie, B.B., McGlasson, W.B. and Wade, N.L. (ed.) (1989). *Postharvest Diseases of Horticultural Produce*, Vol. 1, *Temperate Fruit*. CSIRO Publications, Melbourne, Australia.

Hardenburg, R.E., Watada, A.E. and Wang, C.Y. (1986). *The Commercial Storage of Fruits, Vegetables and Florist and Nursery Stocks*, revised edn. Agriculture Handbook No. 66. US Department of Agriculture, Washington, DC.

Lipton, W.J. and Harvey, J.M. (1977). *Compatibility of Fruits and Vegetables During Transport in Mixed Loads*. Marketing Research Report No. 1070. US Department of Agriculture, Washington, DC.

Peleg, K. (1985). *Product Handling, Packaging and Distribution*, AVI, Westport, CT.

Ryall, A.L. and Pentzer, W.T. (1982). *Handling, Transportation and Storage of Fruits and Vegetables*, Vol. 2, *Fruits and Tree Nuts*, revised edn. AVI, Westport, CT.

Shewfelt, R.L. and Prussia, S.E. (ed.) (1993). *Postharvest Handling. A Systems Approach*. Academic Press, San Diego, CA.

Wade, N.L. and Graham D. (1987). A model to describe the modified atmospheres developed during the storage of fruit in plastic films. *ASEAN Food Journal*, **3** 105–111.

Wade, N.L. and Morris, S.C. (1982). Causes and control of cantaloupe postharvest wastage in Australia. *Plant Disease*, **66**, 549–552.

Watkins, J.B. (1990). *Forced-air Cooling*, 2nd revised edn. Information Series Q188027. Queensland Department of Primary Industries, Brisbane.

Wills, R.B.H., McGlasson, W.B. Graham, D. and Hall, E.G. (1989). *Postharvest. An Introduction to the Physiology and Handling of Fruit and Vegetables*. New South Wales University Press, Kensington, Australia.

4 Production of non-fermented fruit products
P. RUTLEDGE

4.1 Introduction

The extraction of juice from fruit is an ancient art dating from the earliest of records, where wine is often mentioned. Fermentation of fruit juice so the alcohol content preserved the fermented juice was one of the earliest forms of food preservation by the human species. Although fermented beverages are dealt with in another chapter, the extraction of fruit juice must be considered a mature technology. With rapid changes taking place in most technologies during the past century, the manufacture of fruit juice has progressed from the farm or cottage industry into the efficient technology of modern food processing.

Throughout the temperate areas of the world, fruits used for the major quantities of juices are citrus (predominantly orange), pome and grape or vine fruits. Some production of stone fruit and berry juices is carried out but only in small quantities. Pineapple dominates tropical fruit juice production, with highly flavoured fruits such as mango, passionfruit and guava becoming more popular as blending juices.

Methods of extracting fruit juices are dependent upon the structure and edible portion of the fruit. Preservation methods include thermal treatments, freezing, chilling, concentration (drying) and, for some clear juices, fine filtration. Juices may be taken apart by removing volatile flavour components, water, bitterness and acidity and then recombined to produce a consistent product. Fruit-derived drink bases may be manufactured from the remaining fruit material after the juice has been extracted.

4.2 Fruit quality

Fruits used for juice manufacture are often those rejected because of the high specifications for the fresh market, or they may be off-cuts from other fruit processes or fruit which is specifically grown for juicing. Juicing is near the bottom of the fruit usage chain, so care must be taken to ensure that only sound material is used (Figure 4.1). Fruit that is infected with moulds, starting to ferment with yeasts or is rotten is not suitable for juicing and must be removed from the processing line, preferably before

Figure 4.1 Unloading of juice fruit from bulk transport.

washing so that microbial and off-flavour contamination of the juice is prevented.

Fruit maturity is important for optimum flavour, as flavour volatiles are produced near the fully ripe stage. However, if the fruit is allowed to ripen to the stage where senescence begins, the structure of the fruit is degraded and this can cause problems during extraction. For example, pears that are over-ripe form a 'porridge' from which it is almost impossible to extract the juice.

Variety also plays an important part in the quality of the extracted juice. Some varieties are more suitable than others for processing. The classic example of this is the difference between processed Valencia and Navel oranges, with the production of bitter limonin in the Navel juice making it unpalatable. Juices from apples of different varieties may be blended to give the required flavour characteristics.

4.3 Temperate fruit juices

The climate in Europe, northern USA and smaller areas in the southern hemisphere is suitable for growing temperate fruit. Most significant of these temperate fruits are orange, apple and pear, which make up the bulk of processed juice used for non-alcoholic purposes. Smaller volumes of juice from grape, berry and stone fruits are produced.

4.3.1 Orange juice

Valencia and Navel are the two generic varieties of oranges produced in most countries. Other smaller plantings are available for juice, but again these can be separated into the two principle types above. Valencias have a six month season, which is followed by a similar season for Navel oranges. Valencia types produce excellent orange juice that is slightly sour at the start of the season changing through to slightly bland at the end of that season. This is a reflection of the maturity of the fruit and the requirements of processors to extend their season for as long as possible.

Navel orange juice, however, is not only sour at the start of their season but produces bitter processed juice because of the production of limonin. This bitterness is not apparent in the market fruit as it takes time or thermal processing to produce the bitter limonin component. As the season progresses, the amount of bitterness in the juice diminishes and it is also masked by the increasing sweetness of the fruit. Blending preserved Valencia juice with Navel juice during the Navel season does extend the use of bitter Navel juice. The demand for orange juice exceeds the supply of Valencia orange juice alone, making it necessary to juice Navel oranges.

The manufacture of orange juice also includes the production of several by-products. Before the juice is extracted, the fruit is washed and the skin oil is removed by passing the fruit over an oil extraction machine which inflicts hundreds of small cuts on the skin (Figure 4.2). The oil sacs in the surface of the skin are ruptured and the oil is removed by washing. The oil and washwater are centrifuged so that the emulsion is removed from the water, which is returned to the machine. The emulsion can be 'broken' in a high-speed centrifuge to yield the peel oil. In other systems, the oil is removed from the juice after juice extraction, during the finishing operation.

Citrus juice is extracted by either a reaming action (Figure 4.3) or a crushing action, as in FMC extractors. The reaming operation involves cutting the fruit in half followed by the automatic presentation of each half to the reamers to remove the contents. The FMC system uses a cup-shaped set of fingers top and bottom which mesh, as shown in Figure 4.4. As the fruit is crushed between the cups, a small piece of skin is cut from the centre of the bottom cup and the contents are squeezed through the cut

PRODUCTION OF NON-FERMENTED FRUIT PRODUCTS 73

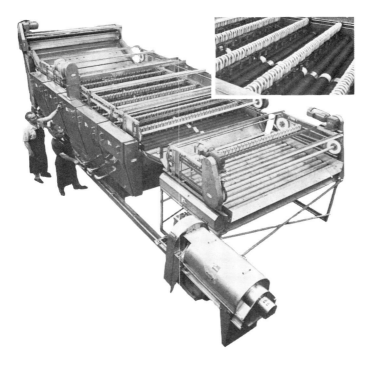

Figure 4.2 Machinery for removing citrus skin oil (courtesy of Brown Machinery Australia Ltd).

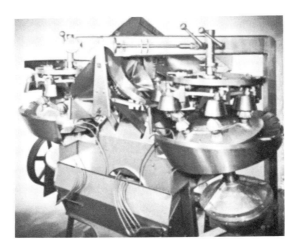

Figure 4.3 Brown reaming type citrus juice extractor (courtesy of Brown Machinery Australia Ltd).

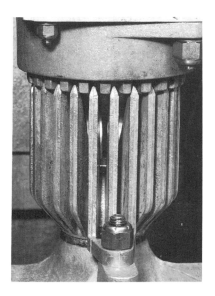 

Figure 4.4 FMC type citrus extractor: left, fingers meshed; right, fingers released.

hole into a tube. The tube is perforated and, via a pumping action, the juice is forced through the perforations into a manifold for removal from the machine. If the skin oil has not been removed prior to extraction, the juice from the crushing operation has a higher oil content than that from the reaming method, assuming correct adjustment of the extractors.

As the citrus juice from the FMC extractors has had to pass through 1 mm diameter holes in the perforated tube, the screening process to remove unwanted material has been partially carried out. Juice cells and excess pulp that remain in the juice are easily removed during the following screening operations. The reaming method results in an excess of white skin lining (albedo), segment walls, cell tissue, pieces of skin and seeds, which must be removed from the juice. A small screw press, paddle or brush finisher using screens with 0.5 mm diameter holes is often used to fine screen the juice. In some plants, the juice is centrifuged to remove pulp, which can then be used in by-products.

Orange juice is heat treated as soon as possible after extraction to inactivate pectinases (enzymes) naturally occurring in the juice. The pectin in the juice is responsible for holding the 'cloud' or suspended solids in the juice. If the pectin is degraded by the pectin esterase, then the attractive cloud in the juice will fall leaving a serum (clear liquid) and a deposit on the bottom of the container. The required heat treatment is 95°C for 30 seconds, which is more severe than that required for microbial stability.

Methods of preservation and marketing of orange juice will depend on the local food laws and customer expectation and acceptance. Orange juice that has been fully pasteurised and received further heat treatments during concentration may undergo detrimental changes in flavour. Customers in the Asian region find thermally induced flavours unpleasant and prefer to purchase orange juice without these flavours.

Orange juice that has been extracted and pasteurised at 65–70°C for a few seconds to inactivate yeasts and moulds will have a shelflife of up to 21 days when stored at 4°C or less. Juice treated in this way does not exhibit undesirable flavours. A shelflife of up to three months at chill temperatures can be attained with orange juice treated with chemical preservatives. Preservatives used, when allowed by law, are typically sulphur dioxide at 100 p.p.m. and a combination of sorbic and benzoic acids at up to 400 p.p.m. Thermal treatment at 65–70°C does not inactivate pectinases, and some separation of the orange juice cloud is evident during storage. As the Valencia season lasts for only half the year, the storage of 'fresh-like flavoured' juice has become an interesting problem. Some manufacturers have found that storage of untreated orange juice at temperatures close to the freezing point extends the storage time without noticeable loss in orange flavour.

Orange juice may be hot filled into cans from the pasteurisation step, inverted and spin cooled to give a high-quality product. Aseptic processing into Tetra-pak or bag-in-the-box systems is common.

Problems associated with sour and bitter juice may be addressed by modifying the juice using adsorbents which remove the unwanted characteristic. These adsorbents are in the form of polymer beads. Johnson and Chandler (1985) investigated systems for removing the bitterness from orange juice and in 1986 they published work on the removal of sourness from citrus juices. Their method is to remove the fruit pulp by centrifuging so the pulp will not clog the bed of beads. The juice is then pumped through the beads and the citric acid or the limonin is removed, depending on the bead used. The pulp is returned to the juice after it has been treated. Beads may be regenerated by passing caustic soda solutions through the bed of beads.

Concentration of orange juice is used to store juice for blending and also for transport to reconstitution plants. The soluble solids of orange juice contain mainly sugar and acid. The percentage of soluble solids is called °Brix (°B) and in orange juice will vary from about 9 to 14°B. Orange juice is passed through multiple effect evaporators under vacuum, which produce concentrate in excess of 65°B. Concentrate is stored chilled or frozen to reduce the non-enzymic browning reactions and attack from *Saccharomyces rouxii*. Highest quality concentrate is produced by removing the flavour volatiles from the fresh juice, at low temperature, in a spinning cone column (see Figure 4.10, below) followed by concentration. The volatiles are returned to the concentrate, which, when diluted, will have the full orange flavour. Other recovery systems remove the volatile components from the distillate during concentration. These volatiles will have undergone some thermally induced changes during this process. When the volatiles are added back to the concentrate, it results in an inferior flavour compared with concentrate using the spinning cone column method.

Other products produced from oranges include pulp wash, where the orange pulp from the finisher or centrifuge is water washed and then refinished. Pulp wash is a water extract of orange pulp and may be used in other products, such as drinks and cordials, where their use is allowed by food regulations. Some manufacturers use a counter-current extractor to extract useful material from the skins which is then suitable for the manufacture of cordial bases. Skins that have been steamed, milled, stone milled and dried also make a comminuted drink base.

4.3.2 Citrus juices

Lemon, lime, grapefruit, tangerine and mandarin are other important citrus fruits. The processing of these fruits into juices is similar to the manufacture of orange juice.

4.3.3 Apple juice

The process used for producing apple juice from apples depends on the volume of production and the amount of capital to be invested. There is a distinction between the processing methods used by a small two or three person operation run by a grower and the larger factory operation run by a company. However, the following processing steps generally apply to both situations.

Before washing, the fruit is inspected to remove rotten or mouldy apples. This is conveniently carried out on an inspection belt in small plants, but with developments in technology such fruit can be automatically removed by scanning systems. Cleaning operations depend on the condition of the fruit entering the factory. If the apples have been hand picked and are free from extraneous matter they will require washing only. If the apples have been machine harvested, which is possible for juice production, then leaves and twigs will require removal by a dry cleaning operation using a high-velocity fan. Washing removes dirt and water-soluble agricultural spray residues and it is important to rinse the apples with fresh water as they are elevated from an immersion tank.

Apples are milled for most extraction systems, but for diffusion extraction they are sliced into 3 mm ripple-cut slices. Hammer mills using

1 to 1.5 cm screens are suitable for producing apple pulp for pressing. This apple pulp is not too finely divided and has the required structure for the pressing operation. To prevent oxidative or enzymic browning, some ascorbic acid is often added at the milling stage, especially if there are likely to be any delays in further processing operations.

Juice may be extracted from the apple pulp or slices by hand-operated rack and cloth pressing, automatic batch pressing, continuous machine pressing, centrifugal separation or by diffusion extraction.

The capital cost of the rack and cloth method of extraction is relatively cheap, but it is very labour intensive. A coarse-weave nylon cloth is spread over the top of a frame placed on a slatted wooden divider. Apple pulp is poured into the cloth until the frame is full. The edges of the cloth are folded over the pulp forming a cloth-bound bed of apple pulp, called a 'cheese' as it resembles the European-style bound cheese. The frame is removed, a divider is placed on the 'cheese' and another 'cheese' is built on top of the first, and so on. A stack of several 'cheeses' is made, depending on the capacity of the press. The stack is placed in a hydraulic press with a collection tray located under the stack. Pressure is gradually applied to the stack until the 'free run' juice is expressed and then pressure is applied up to about 1500 kg in some presses. This process normally takes about 30 minutes. Juice from the rack and cloth press is low in suspended solids and is often processed as cloudy juice without further treatment. Some difficulty is experienced when using old fruit, and a filter aid, such as diatomaceous earth, is put on the cloth before adding the pulp. Juice yield for this method is about 70% of the weight of pulp when using fresh apples but can drop to below 60% with old mushy apples.

There are several designs of automatic batch presses available, with Willmes, Bucher, Bucher-Guyer or Atlas Pacific commonly used for apple pressing. These presses usually need press aid and pectinase added to the apple pulp to aid the pressing operation. The press aid, which is normally shredded paper or rice hulls, forms channels through the press cake for the juice to be expressed. Care must be taken with press aid to ensure that unwanted flavours are not added to the juice.

Automatic batch presses are the most varied in the method used to press the pulp. One uses a large piston to force the pulp into a perforated cylinder, while another has a solid cylinder with perforated flexible nylon tubes between the piston and the end of the cylinder to extract the juice. Another design has the pulp spun onto the walls of a perforated cylinder and an inflatable bag pushes the juice through the cylinder walls. These types of press give juice yields from good apples of 70 to 80%, but the yields from old fruit can drop dramatically. The advantage of these machines is that the labour input is small and they can juice at least 5 tonne/hour of fruit.

Common commercial continuous presses are the screw press and the belt press. The most common is the screw press, which is also used for extracting other kinds of fruit for juice. The screw press consists of a tapered screw surrounded by a close-fitting screen. The end cap of the machine, which resists the exit of the pumice, is held in place by springs or air pressure. The pulp travels along the screw with the pressure being built up by the resistance of the end cap. The juice is expressed through the screen and the press cake is pushed out from the end of the screw when sufficient pressure is applied to overcome the resistance of the end cap. Press aid may be added to the pulp when the apples are old and mushy. A yield of 70 to 75% can be expected when using good crisp fruit. Juice from the screw press often contains excessive suspended solids that have to be removed before a cloudy apple juice can be processed.

Belt presses are continuous presses having two belts through which juice may pass (Swientek, 1985). Apple pulp is spread between the belts which enclose the pulp by coming together before travelling between a series of rollers. The rollers exert pressure on the pulp and juice is expressed through the belts. In some cases, about 1% wood fibre is mixed with the pulp as a press aid. A correctly maintained belt press will yield about the same as a screw press.

Centifugal continuous extraction of pulp is possible using a screening centrifuge such as the Sharples 'Conejector'. In this machine, the pulp is pumped on to the centre of a fast-rotating screen. The juice is forced through the screen and the pumice is also centrifugally removed. The juice is unavoidably aerated during this operation, giving a darker juice than other methods of extraction. The yield is slightly less than continuous pressing and the suspended solids are slightly higher.

Continuous diffusion extraction can be used when the juice is to be manufactured into apple juice concentrate (Figure 4.5). Counter-current extraction on an inclined single or twin screw conveyor is the method normally used. A screen is fitted in the lower end of the machine so liquid can be separated from the solids and removed. Apples are sliced into 2 mm ripple-cut slices and dropped into the lower end of the extractor just above the screen. A small stream of water or distillate from the evaporator is added at the top of the conveyor and runs counter current to the apple slices. The liquid level is adjusted by changing the level of the take-off tube behind the screen. A small stream of the 'juice' is heated in a heat exchanger to about 70°C and sprayed onto the incoming fruit to inactivate enzymes and to reduce the thermal load on the hot water jacket of the conveyor. The slices that exit the top of the conveyor are pressed and the liquid returned to the extractor at the point on the conveyor where the soluble solids are similar. The extraction time is in the order of 1 hour. Casimir (1983) measured yields of soluble solids of 89–91% from Granny Smith apples compared with 68–69% from Bucher-Guyer presses using the same batch of fruit.

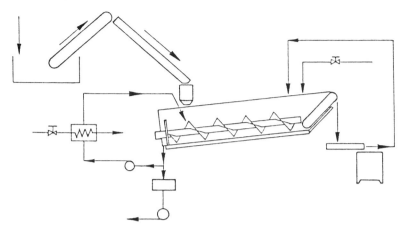

Figure 4.5 Counter-current extraction of apples. 1, Wash tank; 2, slicer; 3, screw conveyor; 4, juice exits through screens; 5, heat exchanger; 6, screw press; 7, returning liquid from screw press; 8, water inlet.

4.3.4 Pear juice

Extraction of pear juice is similar to apple juice. The use of pear juice as a fruit syrup for canned fruit requires it to be clarified, including the brown colour, by amino acid removal. Colour in pear juice may be removed by polymer bead systems (Cornwell and Wrolstad, 1981), although aeration is required to complete the phenolic reactions before removal. If aeration is not carried out, further darkening of the juice will occur on storage. Darkening problems in pear juice concentrate manufactured from treated juice are the result of browning reactions more than phenolics.

4.3.5 Stone fruit juices

Plums, peaches, apricots and cherries are the stone fruits most commonly processed into pulps and juices. With these fruits, the normal crushing operation does not give a reasonable yield because of the structure of the fruit flesh. Plum and cherry juice browns rapidly because of enzymic action once the fruit is broken. To prevent enzymic action, the whole fruit is heated to a temperature that will inactivate the enzymes before it is crushed. This heating is called a hot-break process. This can be carried out using a tomato hot-break machine or by steaming the fruit for several minutes (small plums need 7 minutes) to inactivate the enzymes and to break the structure of the fruit. The fruit is passed through a paddle finisher to remove the skins and seeds from the pulp. The pulp is cooled to about 50°C and then pectinase is added to break the pulp structure further. A decanting centrifuge is used to remove a partially clarified juice from the pulp. Prune plums that are too small for drying give a clarified yield of between 50 and 60% juice containing soluble solids in excess of 20°B. The juice when concentrated to 75°B has a deep maroon colour. Concentrate with these colour characteristics is ideal for blending with lighter coloured juices.

Peach pieces from canning lines, where the peach has been peeled and destoned, that are unsuitable for canning may be pulped mechanically and pasteurised in a scraped-surface heat exchanger. The pulp is concentrated to 30°B without enzyming and aseptically filled into bulk systems for further processing as pie filling or for drying for confectionery.

4.3.6 Berry juices

Berry juices are usually highly flavoured and coloured and as such are ideal for blending with other juices such as clarified apple or pear juice. Importance must be placed upon retaining these characteristics during processing into juice. Berries can be broken between rollers or in mills, but the juice may still be held in the cellular structure of the fruit flesh. Enzymes may be added to break down the structure and the juice recovered in a decanting centrifuge. Heat may be used instead of enzymes, but there might be some loss of flavour and colour as a result.

Preservation of the juice will depend on its ultimate use, but freezing is not uncommon if flavour is important. Thermal processing will allow long-time storage. Concentration of berry juices and subsequent storage of the concentrate, chilled or frozen, is an option if the flavour volatiles are removed before concentration and returned to the concentrate.

Grapes may be considered as berries and some grape juice is clarified for use in blended juices. (The extraction of grapes can be found in the chapter on fermented beverages). The removal of tartrate or the prevention of tartrate crystallisation, sometimes known as 'wine stone', is a problem for long-term storage of the juice. Exchanging sodium for potassium in the juice by ion exchange will change the relatively insoluble potassium hydrogen tartrate for the more soluble sodium salt. Concentrated grape juice can have a sugar precipitate if it is concentrated above about 55°B but this will quickly redissolve on dilution.

4.4 Tropical fruit juices

Tropical fruits require some care in processing if the delicate flavours are to be retained. In some tropical fruits, the acidity normally associated with temperate fruits is lower (with a corresponding rise in pH). Acidification of the juice from these fruits to a pH of about 4 will be required to enable them to be processed at pasteurisation temperatures.

4.4.1 Pineapple juice

Pineapple juice is often recovered from the ejected skins and cores from the Janaka machines which cut the pineapple in preparation for canning. Other sources of fruit for juice are small pineapples, physically damaged fruit that is unsuitable for canning and off-cuts from the canning line. This waste material is utilised by using a screw press to recover the juice. In some cases, a finisher with a fine screen or centrifuge is used to ensure a light opalescent juice is collected. Pineapple juice is a high-acid product that is preserved in a similar manner to normal fruit juices.

4.4.2 Papaya purée

Papaya is one of the bland tropical fruits that may require acidification to ensure a safe product when processed by normal methods. Fruit is chopped into pieces large enough to fit into a paddle finisher where the seeds and skin are removed from the flesh. Only ripe fruit should be used as the finisher will not extract efficiently if hard pieces of fruit are fed into the machine. Pulp is blended with citric acid to give the required pH, which should be below 4.2 but is often as low as 3.5. Blended pulp is then fed through a scraped-surface heat exchanger to heat the pulp to 94°C. It is held for 2 minutes to ensure microbial stability followed by cooling in a scraped-surface heat exchanger prior to aseptic packaging (Morris, 1982). The heated pulp may also be hot filled into cans, inverted and cooled. In some countries, the papaya fruit is comparatively high in nitrates, which will corrode tinplate cans very quickly and unduly shorten the shelflife of this product.

Freezing papaya pulp has been used by some American processors for the preservation of unacidified pulp. This is a relatively expensive method to use but necessary if unacidified pulp is required by the customer. The other alternative would be to give a botulinum-type thermal process, but the pulp would be badly damaged by the heat treatment, with the loss of the delicate flavour.

Treatment of papaya pulp with pectinases will reduce the pulp to a solid and a serum after separation in a decanting centrifuge. The serum may be pasteurised if acidified and concentrated, but this product does not have the appeal of the pulp.

4.4.3 Mango pulp

Mango pulp is a popular blending material for many types of mixed fruit juice product. The mango has a large stone, stringy sometimes fibrous flesh and a skin that if included in the pulp, has a characteristic terpene flavour which can be unpleasant. Pulp is extracted by steaming the fruit for 2 to 3 minutes, followed by a rough chop to break the fruit but not break the stone. Chopped mango is extracted with a paddle finisher using a screen fine enough to eliminate the fibrous material in the flesh. This produces a pulp that may need acidification for preservation, like the papaya pulp.

Mango is also extracted using counter-current systems, which have been described in the extraction of apple juice (section 4.3.3). The mangoes are washed and then squashed to break the fruit without breaking the stone. The squashed fruit is dropped into the lower end of a counter-current extractor (Figure 4.6) with water at 65°C running countercurrent to the fruit. Reheated 'juice' is sprayed onto the incoming fruit to inactivate enzymes and reduce the thermal load on the system. A wiper is required on the bottom screen to prevent the build up of fibrous material, which would eventually block the screen. The 'juice' from the extractor has been diluted up to 10% by water and required concentration to bring it back to the original strength. Mango from the counter-current extractor may be concentrated to 30°C without enzyme treatment. Excellent quality, light

Figure 4.6 Extraction of mango using a counter-current extractor.

yellow-orange-coloured mango juice concentrate is produced from counter-current extracted mangoes.

4.4.4 Passionfruit juice

Passionfruit has two main varieties, the purple and the yellow types. The purple variety has a superior flavour to the more prolific and larger yellow variety. Passionfruit has a delightful flavour (Whitfield and Last, 1986) which is used for flavouring drinks or for blending with other juices.

Extraction of passionfruit pulp is accomplished by dropping the washed fruit between two rotating converging cones. Fruit is caught in the nip of the cones and burst as the cones rotate towards the bottom of the machine where the clearance is reduced to thickness of the skin of the fruit (see Figure 4.7). The skins carry through in the cones and the pulp drops into a finisher that removes pieces of skin and, depending upon the finisher screen size, also the seeds.

Figure 4.7 Twin-cone passionfruit extractor with the top cover removed.

Passionfruit juice has a higher starch content compared with other fruit juices. If the starch is not removed, thermal processing produces a highly viscous product that is ideal for some uses. However, the starch may be removed in a decanting centrifuge leaving a juice that can be concentrated to about 50°B. Casimir *et al.* (1981) have discussed the types of evaporation equipment suitable for concentrating passionfruit juice. The important requirement for the equipment is a very short residence time in the evaporator to prevent thermal damage, which will destroy the flavour of the juice.

High-quality passionfruit juice is often frozen in preference to thermal processing so the flavour remains intact. Freezing is an expensive option as the cost of storage increases with time, but the flavour of passionfruit is its main attribute and worth preserving. Short-time pasteurisation processes from spin-cooking cans or for aseptic processing also produce a high-quality passionfruit juice.

4.4.5 *Guava pulp*

Guavas are similar to pears for processing as both fruits have stone cells (sclereids) in the flesh. Stone cells in juices give an unpleasant 'gritty' mouthfeel that is unacceptable. The guavas are washed and chopped for presentation to a paddle finisher fitted with a screen containing holes of about 1 mm diameter. A second finisher with a finer screen of about 0.5 mm holes is used to remove stone cells. An alternative method to remove the stone cells is to grind the juice in a stone mill; this will make the stone cell small enough not to have an unpleasant mouthfeel in the juice. Stone cells may also be removed by centrifuging the juice, but this may also remove some of the wanted pulp material.

Guava pulp may be preserved by freezing without pasteurisation. Aseptic processing is also a popular method of preserving the pulp (Figure 4.8), which requires a temperature of 95°C for 30 seconds for microbial stability.

Pectinase-treated guava pulp will produce a clear guava juice after pressing or centrifuging in a decanter centrifuge. If a very clear juice is required, then filtration will give a polished fruit product. This juice can be used for clear jellies or blending with other clarified juices such as apple juice. Pectinase-treated guava pulp may be concentrated to about 30°B and the clarified juice to 70°B.

4.5 Clarification of fruit juices

The initial stage of juice clarification is the removal of excess pulp, which can be carried out by centrifugation in a decanting centrifuge or by

Figure 4.8 Aseptic processing of fruit pulp (courtesy of Tetra Pak).

finishers with fine screens. Juice is clarified by removing pectin, starch, gums, proteins, polyphenolics, metal cations and lipids, which may cause hazes before or after preservation. Specialised enzyme preparations are available for specific fruit juices; for example, a mixture of pectinase and amylase might be used for early season apples. If an enzyme is used in the mash (Anon, 1985), then this mixture might include some arabinase.

The traditional method of fining is to heat the juice to the required temperature and stir in the enzyme or enzyme mixture, wait until the enzyme has acted and then add agents that will bring down tannins and other unwanted material (Heatherbell, 1976a,b). These agents are gelatine, bentonite and silica-sol and flocculation usually occurs rapidly after their addition. This is carried out at temperatures near ambient to get good flocculation; however, Grampp (1977) describes a system for hot clarification that is faster than the traditional method. The juice is decanted or racked from the lees and filtered using a filterpress or a diatomaceous earth filter such as a rotary vacuum filter.

Control of the traditional method of fining was often difficult, with occasional unexplained additional flocculation occurring after filtration. This often occurred when pectin tests, using acidified alcohol, were carried out by inexperienced personnel who were not able to distinguish a fine pectin haze. Membrane filter systems have become a practical reality since the cross-flow system started to be used industrially in the 1970s (Pepper, 1987) and these have eliminated many of the haze problems.

Ultrafiltration of fruit juice has become the industry standard because of the simplicity and effectiveness of the system. The juice is enzymed to reduce clogging of the filter and then passed across the filter under pressure. The pores in the filter are small enough to hold back tannins and other compounds that cause haze formation in the clarified juice. Yields of juice from ultrafiltration are typically in the 95 to 97% range, compared with 90 to 93% from the traditional method. Chemicals used in the traditional fining are no longer required, although the ultrafilter cartridges have to be replaced.

4.6 Methods of preservation

Prior to any preservation steps, the juice should be deaerated to remove dissolved or entrapped O_2 which will react with ascorbic acid and also darken the juice. This entails spraying the juice into a chamber that has been partially evacuated and then pumping the juice from the bottom of the chamber.

4.6.1 Thermal treatment of the juice

As fruit juice is a liquid food, the easiest method of heating is by heat exchanger. There are several types of heat exchanger and the choice of type will depend on the amount of insoluble solids in the fruit juice. Clear juices and those with only a small amount of insoluble solids can be heated in a plate heat exchanger, preferably with a regeneration section.

Spiroflo type heat exchangers have more widely separated internal surfaces than the plate type and, therefore, will accept juices with a higher proportion of solids. These can be used for quite viscous products and could be used for single strength pulp. Shell in-tube heat exchangers are similar to Spiroflo but do not have the advantage of turbulent flow.

Scraped-surface heat exchangers are used for pulp with high levels of solids and for concentrated pulps such as 30°B peach pulp concentrate. Scraped-surface heat exchangers have a jacketed drum that can be heated or cooled. The product is wiped with a scraper along and around the inside of the drum. This equipment will quickly heat or cool products with high viscosity and poor thermal characteristics.

4.6.2 Canning

With the exception of some tropical fruits, fruit juices and pulps are acid products with a pH of less than 4.2 and often in the 3.5 to 4.0 range. To inactivate microbial growth at these pHs, the juice requires heating to 80 to 93°C for a few seconds only. Plain tinplate cans are used for fruit juice as the tin has a reducing effect on the juice. Juice that has darkened as a result of oxidation will return to its natural colour under the reducing influence of the tin–acid reaction of corroding tinplate cans. This is the major difference between packing juice in cans and in plastic or plastic–cardboard containers. Juice packed in plastic will darken with time.

Preserving juice in tinplate cans can be carried out by hot filling, closing and cooling the cans at speeds in excess of 500 cans/minute. The juice is heated in a heat exchanger to the required temperature, filled into the can, steam flow closed, inverted to pasteurise the end and, ideally, spin cooled under water sprays. The ideal rate of spin for 74 mm diameter cans is 180 r.p.m. for single-strength thin fruit juice.

Spin cooking juice in cans that have been vacuum closed is another efficient method for canning juice. Vacuum closers are slow, up to 200 cans/minute, but the cans are thermally processed by spinning them in atmospheric steam and then cooling them under water sprays. Processing time for orange juice is between 60 and 90 seconds and cooling takes about 45 seconds. Canned juices with superior quality are produced from spin cooling or spin cooking–cooling.

Cooking cans in boiling water is a further method, but the product is usually heat affected and the market place will not accept this poor-quality product.

4.6.3 Aseptic processing

Technically, aseptic processing is a thermal process like canning where the product is pasteurised and then filled without contamination into sterilised containers and hermetically sealed. The Tetra Pak system seen in Figure 4.9 uses a heat exchanger to pasteurise and cool the juice, which is filled into a cardboard–aluminium–plastic laminated box. The box material enters the machine as a sheet that is sterilised with hot peroxide solution, formed into a tube, filled, sealed, cut and then folded into the box shape.

Other aseptic systems are designed for filling bags that will fit into 200 litre drums or into boxes about the size of 0.5 tonne fruit bins. It is possible to fill road and rail tankers using the aseptic system. In the case of tankers, they are steam cleaned and then filled with sterile air as they cool before filling with product.

Figure 4.9 Tetra Pak aseptic packaging machinery (courtesy of Tetra Pak).

4.6.4 Bottling

Bottling fruit juice is similar to hot filling cans but uses glass bottles. The bottles pass under hot-water sprays after capping to pasteurise the cap, or the cap can be put on the bottle under steam. Cooling is achieved by passing the bottles under water sprays, using warm water initially to prevent breakage through thermal shock on the glass.

Plastic containers are more difficult to fill because of the thermal instability of the plastic. Bottles made from polypropylene will withstand the temperature of normal hot filling but the material is unsuitable for storing fruit juice because of its high permeability to O_2. Early plastics used for storing fruit juices were unstable above 72°C. These bottles may be used if they can be held at 70°C for 10 minutes after filling. Juice is heated in a plate heat exchanger to 85°C and then cooled in a regeneration section to between 71 and 72°C before filling the bottles. The bottles are squeezed during filling to remove any headspace and are then held for 10 minutes at 70°C before water cooling. Polyester materials such as polyester terephthalate (PET) can be hot filled like glass bottles.

Shelflife of plastic containers with fruit juice is in the order of six months before the permeation of O_2 through the plastic either darkens the juice or oxidises the constituents such as ascorbic acid. Although plastic containers have technical difficulties with filling and pasteurisation, they are favoured by marketing because of the different shapes available and the good presentation of the product.

4.6.5 Chemical preservatives

Juice treated with chemical preservatives such as sulphur dioxide, sorbic and benzoic acids will have a shelflife of many months at 4°C. In some countries, the use of preservatives in juices is banned. The levels required to preserve juice at 4°C are about 100 p.p.m. sulphur dioxide (usually added as sodium metabisulphite) and 400 p.p.m. of a combination of sorbic and benzoic acids (usually added as the sodium or potassium salts).

4.6.6 Freezing

Freezing of juice is carried out where the pH of the pulp is high or the preservation of a particular flavour component is important. The juice is chilled in a heat exchanger and then filled into plastic containers that are frozen in a blast freezer using air temperatures of about −40°C. The juice should be frozen to −18°C and stored at that temperature.

4.6.7 Filtration sterilisation

Membrane filters are able to filter clarified juices such as apple and grape juice so finely that the yeasts and moulds that normally spoil juices are eliminated. Mulvany (1966) describes some early filter systems. Grape juice stored for future wine fermentation is commonly sterile filtered and stored in sterile tanks. Sterile filtering replaces the treatment of the grape juice with sulphur dioxide and eliminates the problem associate with removing the preservative before fermentation can take place.

4.7 Concentration of fruit juice

4.7.1 Essence recovery

Fruit is prized for the esters and flavours that are produced during ripening. Importance must be placed on the retention of these flavours during processing of the fruit into juice and through any subsequent treatments. Most of these flavour components have a boiling point less than that of water and are considered to be 'volatile' during concentration. Volatile flavours are lost into the distillate stream during concentration. In many of the older concentrators there was an essence recovery column for the distillate, which removed the volatile flavours for inclusion into the concentrate. These volatiles had received the heat treatment given to the juice during concentration and then another heat treatment to remove them from the distillate. Heat treatments for concentration are normally carried out under vacuum and, therefore, the temperatures are not high, but some degradation of the flavour does occur through each cycle.

A better method is to remove the volatiles prior to concentration and a simple way to do this is in a spinning cone column. The column consists of a series of cones with every second cone attached to a spinning shaft in the centre and every other cone attached to the wall of the column, as in Figure 4.10. The juice passes through a heat exchanger to warm the juice to just below its boiling point at the vacuum applied in the column. The temperature used is normally between 40 and 50°C. The juice enters the column at the top and runs down the first cone by gravity. It falls into the gap between the cone and the shaft onto the spinning cone. The juice is spun to the wall under the centrifugal force on the spining cone and so on down the column. This presents a large surface area to the gaseous phase which enters the bottom of the column and travels countercurrent to the juice. The gaseous phase may be air, nitrogen, or steam under vacuum. The volatiles are picked up by the gaseous phase and removed from the top of the column for condensation and removal. The juice passes out of the bottom of the column and is removed by a bottoms pump (Figure 4.11).

Resident time for juice in the spinning cone column is in the order of a minute so little thermal damage is done to the juice, which is then pumped to the evaporator for concentration.

Other types of column are available for essence recovery before concentration. These columns act like an effect in a multiple-effect evaporator, as shown in Figure 4.12. Juice is heated in a heat exchanger

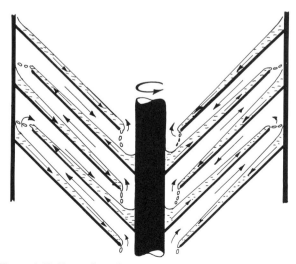

Figure 4.10 Extraction of volatiles by the spinning cone column.

Figure 4.11 A commercial spinning cone installation.

and then flashed into a cyclone with the vapour entering a fractionating column. The juice is pumped from the bottom of the cyclone into the first stage of the evaporator as it retains some heat from the flashing process. The vapour passes up the column and into a condenser to remove the essence fraction.

4.7.2 Concentration

Concentration of fruit juice involves boiling the juice under vacuum and removing the steam by condensation. There are several types of concentrator but the most efficient are the multiple-effect evaporators. Pollard and Beech (1966) outline methods of vacuum concentrating fruit juices, but since that time multistage, multieffect, thermal-accelerated-short-time-evaporators (TASTE) have evolved. These evaporators are described by Nagy *et al.* (1993). A stage and effect has a preheater, distribution cone, tube bundle and liquid vapour separator. The tube bundle has juice travelling down the inside of the tubes while the vapour from the previous stage is travelling up the outside of the tubes. A typical layout is shown in Figure 4.12.

Centrifical evaporators like the Alfa-Laval Centritherm have very short holdup times of about a second. Conical design of the heat-transfer surface in the Centritherm increases the g forces on the concentrate as the viscosity

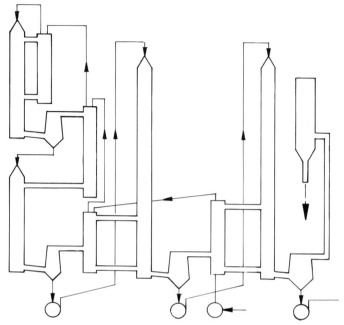

Figure 4.12 Multiple-effect evaporator. 1, heaters; 2, entry of product; 3, exit of product; 4, cyclone; 5, tube-bundle heat exchanger; 6, condenser.

increases through concentration. Although these machines are only a single effect and, therefore, expensive to operate compared with multiple-effect evaporators, they produce excellent quality juice concentrate (Robe, 1983).

Juices with delicate flavours that are affected by the heat treatment during concentration may be freeze-concentrated. As ice forms in the juice it may be removed, leaving behind a concentrate. This process is explained by Tannous and Lawn (1981) but they also found that on freezing apple juice some constituents apart from water were removed with the ice.

Combinations of ultra-filtration and reverse osmosis have been used to concentrate fruit juices without heating (Mans, 1988). Although membrane technology has been available for some time, it has not gained the degree of acceptance by the industry as other forms of juice concentration have done.

4.8 Products derived from fruit juice

Fruit juice in most countries is defined in 100% fruit content. Products derived from fruit such as nectar, fruit juice drinks, cordials and carbonated beverages have definitions as to their fruit content that will depend on national food regulations. Australian regulations, for example, have definitions for diluted fruit juice that contain at least 50%, 35% and 5% fruit juice. Also for a cordial, syrup or topping, the fruit juice content must be at least 25% and the soluble solids a minimum of 25°B. Therefore, fruit nectar under these regulations is a fruit juice syrup containing at least 25% fruit juice. Regulations will vary from country to country, so fruit juice drink products will also vary.

4.8.1 Fruit juice drink

Fruit juice drink will contain fruit juice that has been diluted with a sugar–acid syrup. In most cases, the acidity is in the order of 1% and the soluble solids will be between 10 and 12%. As the pH is between 3.0 and 4.0, the drink can be preserved in a similar manner to fruit juice.

4.8.2 Fruit nectars

Fruit nectars are a mixture of fruit juice, water and sugar. The soluble solids of nectars can be between 25 and 50% depending on country and customers. These products are acidic and can be preserved as for fruit juices, with an allowance for the increased viscosity. The juice used in nectars is usually a full pulp juice and additional pulp may be added in some cases.

4.8.3 Carbonated beverages

Carbonated beverages may or may not contain fruit juice. In general the carbonated beverages which are sold as soft drinks are a mixture of flavoured, coloured and acidified syrup, which is stabilised with benzoic acid (sodium benzoate). This syrup is diluted with soda water and filled into bottles or plastic containers under cool conditions to preserve the carbonation before capping. Adcock (1986) explains the conditions for carbonation and the removal of air from the carbonator. Figure 4.13 shows the bottle pressure required to achieve various volumes of carbonation with CO_2.

Carbonated fruit juice is manufactured in some countries. The fruit juice is cooled and sparged with CO_2 before entering the carbonator where the juice is carbonated under pressure with CO_2. Carbonated juice is filled into glass bottles, capped and passed through a bottle pasteuriser so the product reaches 70°C, which is held for 10 minutes. Bottles are cooled in the pasteuriser under water sprays. The bottles for this process must be strong enough to withstand the pressure of the carbonation at 70°C. In practice,

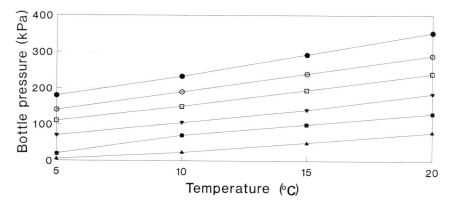

Figure 4.13 Carbonation pressures. ──▲── 1.5 volumes; ──■── 2 volumes; ──▼── 2.5 volumes; ──□── 3 volumes; ──○── 3.5 volumes; ──●── 4 volumes.

there is some breakage in the pasteuriser because of variation in the wall thickness of bottles.

4.9 Adulteration of fruit juice

For various reasons, the manufacturers of modern fruit juices have adopted systems where fruit juice is taken apart, modified and then put back together again. Concentration is a typical example of this, for example, citrus fruit juice has had the pulp, skin oil and essence removed prior to concentration. It could have been debittered or deacidified and a skin extract could have been made. There is a great temptation for a manufacturer with all these fractions at hand to manufacture some pulp wash, add some comminuted pulp for body, some of the essence, sugar, acid, juice and some water to make a product similar to fruit juice.

Manufacture of 'artificial' fruit juices prompted the importing countries in Europe and Germany in particular to look into ways of detecting adulteration of fruit juice. Germany's solution was to look at many detectable attributes of the particular juice and determine RSK values. RSK stands for Richtwert (guide value) Schwankungsbreite (range) and Kennzahl (reference number) (VdF, 1987).

America also has problems with adulteration of fruit juices, and Brause (1993) recounts the Food and Drug Administration's success in convicting companies by using different analytical determinations. Referring to the American situation, he states "Despite all the less than honest people who have been caught at it, adulteration continues to occur at about a 10% incidence."

The Campden Food and Drink Research Association in the UK has developed a near infrared (NIR) spectroscopy system for verifying fruit juices (Scotter et al., 1992). Using NIR, a wide variety of juice characteristics can be predicted: for example, country of origin, several kinds of adulteration, the type of process that the sample came from as well as the fruit variety from which the juice was derived. The spectral database currently contains 1000 samples from seven species, and over half are orange juices. In a blind test of 46 samples, including authentic and adulterated samples of seven juice types, 85% were correctly identified. The results prove NIR to be a valuable screening technique for fruit juice verification, and, as the database expands, the predictive ability of the technique will become more accurate.

Other countries also have their systems for determining whether fruit juice has been adulterated. There does not appear to be any universal foolproof system, as this would have been adopted by all countries. Unfortunately, while the rewards for adulteration are so high the unscrupulous few will attempt it without regard for the good name of the industry.

References

Adcock, G.I. (1968). *The Jusfrute Book – A Text Book of Soft Drink Manufacture.* Jusfrute Ltd, Gosford, NSW, Australia.
Brause, A.R. (1993). Detection of juice adulteration. *Journal of the Association of Food and Drug Officials,* **57**(4), 6–25.
Casimir, D.J., Kefford, J.F. and Whitfield, F.B. (1981). Technology and flavour chemistry of passionfruit juices and concentrates. *Advances in Food Research,* **27**, 243.
Casimir, D.J., (1983). Countercurrent extraction of soluble solids from foods. *CSIRO Food Research Quarterly,* **43**, 38–43.
Cornwell, C.J. and Wrolstad, R.E. (1981). Causes of browning in pear juice concentrate during storage. *Journal of Food Science,* **46**, 515–518.
Grampp, E. (1977). Hot clarification process improves production of apple juice concentrate. *Food Technology,* **Nov**, 38.
Harrison, P.S. (1970). Reverse osmosis and its applications to the food industry. *Food Trade Review,* 33.
Heatherbell, D.A. (1976a). Haze and sediment formation in clarified apple juice and apple wine – I: the role of pectin and starch. *Food Technology in New Zealand,* **May**, 9.
Heatherbell, D.A. (1976b). Haze and sediment formation in apple clarified apple juice and apple wine – II: the role of polyvalent cations, polyphenolics and proteins. *Food Technology in New Zealand,* **June**, 17.
Johnson, R.L. and Chandler, B.V. (1985). Ion exchange and absorbant resins for removal of acids and bitter principles from citrus juices. *Journal of the Science of Food and Agriculture,* **36**, 480–484.
Johnson, R.L. and Chandler, B.V. (1986). Debittering and de-acidification of fruit juices. *Food Technology in Australia,* **38**, 294–297.
Mans, J. (1988). Next generation membrane systems find new applications. *Preparared Foods,* **April**, 69–72.

Morris, C.E. (1982). Aseptic papaya puree. *Food Engineering*, **April**, 84.
Mulvany, J. (1966). Filtration-sterilisation of beverages. *Process Biochemistry*, **Dec**, 470.
Nagy, S., Chen, C.S. and Shaw, P.E. (1993). *Fruit Juice Processing Technology*. Agscience, Auburndale, FL.
Pepper, D. (1987). From cloudy to clear. *Food*, **Oct**, 65.
Pollard, A. and Beech, F.W. (1966). Vacuum concentration of fruit juices. *Process Biochemistry*, **July**, 229–233, 238.
Robe, K. (1983). Evaporator concentrates juices to 70°B in single pass vs. 2 to 3 passes before. *Food Processing*, 92–94.
Scotter, C.N.G., Hall, M.N., Day, L. and Evans, D.G. (1992). The authentication of orange juice and other fruit juices. In *Near Infra-Red Spectroscopy*, pp. 309–314. (ed. K.I. Hildrum *et al.*) Ellis Horwood, Chichester.
Swientek, R.J. (1985). Expression-type belt press delivers high juice yields. *Food Processing*, **April**, 110.
Tannous, R.I. and Lawn, A.K. (1981). Effects of freeze-concentration on chemical and sensory qualities of apple juice. *Journal of Food Science Technology*, **18**, 27–29.
VdF (1987). *RSK values. The Complete Manual*. VdF Verband der deutschen Fruchtsaftindustrie e.V., Bonn.
Whitfield, F.B. and Last, J.H. (1986). The flavour of passionfruit – a review. *Progress in Essential Oil Research*. Walter de Gruyter, Berlin.

5 Cider, perry, fruit wines and other alcoholic fruit beverages
B. JARVIS

5.1 Introduction

The fermentation of fruit to produce wines, as well as the brewing of beer, is recorded in ancient Egyptian and Greek writings. Although production was based largely on the fermentation of grape juice, there is no doubt that fermentation of fruits other than grape had been practised widely, although because of the lower alcohol content such wines did not store well. Over the years the production of grape wine became dominant, except in those areas where cultivation of vines was limited by climatic conditions – in such areas wine was produced by the fermentation of juice from other fruits.

Only the product of fermentation of grape juice can be ascribed the unqualified name of 'wine'. Fermented fruit juices are, therefore, given the generic term 'fruit wines' or, in some places, 'country wines'. Techniques for their production closely resemble those used for fermentation of grape wine, cider and perry (which are special categories of fruit wines).

The development of distillation practices to produce stable spiritous beverages led to the development of distilled products from wines of all types. In addition, some distilled products were subsequently flavoured by percolation through macerated fruit or by steeping the fruit in a distilled spirit. Further variations have included the blending of fruit juices, sugar and other ingredients, such as herbs and vegetative plant extracts, with distilled spirits to produce liqueurs and aperitifs.

This chapter considers the characteristics of a range of fruit-based alcoholic beverages (other than grape wine, on which many books have been written) and the manner of their production. The principles of the production of cider are described in some detail. Shorter accounts are given of the production and properties of perry, fruit wines, fruit liqueurs and fruit spirits. Fruits that are used to produce alcoholic fruit-based beverages include both cultivated and 'wild' species. Some characteristics of alcoholic fruit-based products are summarised in Table 5.1.

Table 5.1 Characteristics of fruit-based alcoholic beverages

Type	Fruit source	Alcohol content (g/100 ml)
Cider	Apples	1.2–8.4
Perry	Pears	1.2–8.4
Fruit wines	Various	6–14
Fortified fruit wines	Various	8–20
Fruit liqueurs	Various	20–30
Fruit spirits	Various	35–40

5.2 Cider

Cider (*syn*: cyder, hard cider, apple jack, cidre, apfelwein) is the alcoholic product of fermentation of apple juice and differs from the non-alcoholic product sold in the USA. In recent years, cider has become an increasingly important commercial product in both domestic and export trade, although over 60% of the European cider is produced in England. Both commercial and 'farmhouse' ciders are produced in many countries. This account reviews the current state of knowledge of cider fermentation and the factors that affect its production.

5.2.1 A brief history

The fermentation of apple juice to produce an alcoholic beverage is believed to have been practised for over 2000 years. The fermented juice of apples is recorded as being a common drink at the time of the Roman invasion of England in 55 BC. Celtic mythology revered the 'sacred apple' and in his famous history (AD 77) Pliny the Elder refers to a drink made from the juice of the apple. Cider was drunk throughout Europe in the third century AD and in the fourth century, St Jerome used the term *sicera* (from whence the name cider was possibly derived) to describe drinks made from apples.

Cider was reputedly a more popular drink than beer in the eleventh and twelfth centuries in Europe. Traditionally, cider has been produced throughout the temperate regions of the world where apple trees flourish; such localities included the northern coastal area of Spain; France, especially Normandy and Brittany; Belgium; Switzerland and Germany; various regions of England; Finland, Sweden; and more recently, Canada, the USA, Australia and New Zealand.

There are references to cider in many writings from the Middle Ages. The popularity of cider in fourteenth century England was such that William of Shoreham reflected the Church's concern for the niceties of sacramental rites by stating that 'young children were not to be baptised in cider'! William Langdon in *Piers Plowman* and Shakespeare in *A Midsummer Night's Dream* refer to the consumption of cider; Daniel Defoe observed that Hereford people 'boaft the richeft cider in all Britain' and Samuel Pepys noted in his Diary that on 1st May 1666 he 'drank a cup of Syder'. From the seventeenth century onwards, cider was praised in numerous poems and other literary works as an aid to good cheer and a homely cure for almost every ailment known to man.

Up to the twentieth century, cider was a popular rural drink, cheaper than beer and often more potent at *c.* 7% alcohol by volume (ABV). In England, farm workers often had part of their wages paid as truck (i.e. in kind) and every farmer would make his own cider for consumption by his workers and his own family and guests. Most farms in the West of England and the West Country had their own cider press, or used the services of a travelling cider press that was hauled by horse from farm to farm.

Commercial cider production commenced during the latter part of the nineteenth century in England, although a few farms had sold cider from early in the eighteenth century. Cider production in England was estimated at 55 million gallons in 1900. By 1920 the level of cider production had decreased significantly; although some 16 million gallons were still produced on farms, only 5 million gallons came from factory producers. In the early 1980s cider sales had risen to over 60 million gallons per year and in the 12 months to December 1994 a total of 92 million gallons was sold in the UK.

The growth in cider production in the UK, which had been inhibited by the imposition of Excise Duty in 1976, results both from the introduction of new concepts in cider products and from changes in drinking habits, especially of younger adults. In 1994, it was possible to obtain cider on draught (both keg and cask conditioned), prepackaged in glass and PET bottles and in cans. Products range in alcoholic strength from less than 0.5% ABV to 8.4% ABV, and in sweetness from totally dry (i.e. no residual sweetness) to sweet. In addition, cider products now range from 'white' ciders (fermented from decolorized apple juice or produced by decolorization of fermented cider) to 'black' cider (a blend of cider and fermented malted barley). Cider made from apples of a single crop may be sold as a defined year Vintage; whilst others may be made from a single apple cultivar, e.g. Kingston Black.

5.2.2 *Cider and culinary apples*

Traditional European ciders are made from the juice of cider apples, believed to have been imported into Britain by the Normans, although it is recorded in Gaulmier's *Traite du Sidre* (1573) that a Spaniard named Dursus de l'Etre brought apple trees into France in 1486 (Jarvis, 1993a).

Traditional cider apples are of four main types: bitter sweet, low in

Table 5.2 Characteristics of selected cider apple varieties

Type	Typical varieties	Typical composition	
		Acidity (g/100 ml)	Tannin (g/100 ml)
Sweet	Sweet Coppin	0.20	0.14
	Taylor's Sweet	0.18	0.14
	Court Royal	0.21	0.11
Bitter Sweet	Dabinett	0.18	0.29
	Michelin	0.25	0.23
	Yarlington Mill	0.22	0.32
	Néhou	0.17	0.60
	Médaille D'Or	0.27	0.64
Bitter Sharp	Kingston Black	0.58	0.19
	Breakwell's Seedling	0.64	0.23
	Bulmer's Foxwhelp	1.91	0.22
Sharp	Brown's Apple	0.72	0.13
	Reinette O'bry	0.63	0.13
	Frederick	1.02	0.09

acidity but high in tannin; bitter sharp, high acidity and high tannin; sharp, high acidity but low tannin; and sweet varieties, low acidity and low tannin. Details of some typical cultivars and the composition of their juices are given in Table 5.2.

In some areas, especially Kent, Suffolk and Norfolk in the UK, and in Germany and South Africa, most cider is made primarily from culinary apple varieties such as the Granny Smith and the Bramley. Whilst such varieties contain low levels of tannin, the juices tend to have a higher acidity than do the more traditional cider apples. In Germany, increased astringency is developed during the production of *Frankfurter apfelwein* by suspending bags containing berries of the *Sorbus* spp. in the fermentation vats to extract the astringent components. Blends of bittersweet and culinary juices are frequently used to develop particular flavour profiles in commercial cider blends.

Cider orchards. The traditional farm orchard of standard trees still exists and, although generally declining in acreage, many hundreds of such orchards provide both an apple crop and grazing for livestock. Modern cider apple orchards are largely intensive bush orchards with trees frequently planted as closely as 2.5 metres apart in rows some 5–6 metres apart. After planting, staking and protecting using a wire guard against rabbits, the grass under the trees is treated with a suitable herbicide to reduce competition for nutrients and water. Some growers cover the herbicide strips with a mulch of apple pomace, woodchips, straw or other suitable material to ensure maximum retention of moisture. Although it has become standard practice to retain the herbicide strip into the productive years of an orchard, this practice is currently the subject of research to assess whether a sward of slow-growing grass varieties may be better in relation to the quality of harvested fruit.

During the growing season, it may be necessary to spray the trees against pests (e.g. red spider mite), mildew, canker, apple scab and other conditions. In addition, in years of heavy potential cropping, chemical thinning (e.g. using carbaryl) is frequently recommended. The objective is to reduce stress on the tree in order to minimize the risk of biennialism, a condition that has occurred significantly in English cider fruit orchards over the past 20 years.

Harvesting and pressing. Traditionally the apples, falling naturally or shaken by hand from the trees, would be raked into piles (Figure 5.1) or filled into sacks that were stored under the trees until the fruit was in a suitably ripe condition to be milled and pressed. In modern intensive bush orchards (Figure 5.2), it is normal to shake the trees mechanically in order to cause the fruit to fall. Such shaking does not harm the tree and permits fruits to be harvested mechanically immediately after falling, so reducing the risk of the rots that may occur if fruit are left for any length of time on a herbicide strip. The collected fruit will normally be washed mechanically in

Figure 5.1 Traditional harvesting of cider apples from standard trees. Apples were shaken to the floor of the orchard where they were gathered by hand into baskets and then tipped into piles to mature.

Figure 5.2 A modern bush cider apple orchard showing the herbicide strip along the row under the tree canopy.

the orchard and then transferred by road to the cider mill where, after weighing, the fruit is tipped into a fruit canal (Figure 5.3) or onto a concrete pad prior to further washing, milling and pressing.

Traditionally the fruit was crushed between stone rollers and the fruit mash was then pressed in a screw press after being built into a 'cheese' consisting of layers of fruit in hessian cloths held on ash slats (Figure 5.4). In modern processing, the fruit is milled using either a hammer or knife mill; the mash is then transferred into a batch press such as a Bücher-Guyer hydraulic ram press or a continuous belt press such as a Belmer (Figures 5.5 and 5.6). The pomace from the primary pressing may be extracted with warm water and repressed to give a secondary juice with a lower sugar content; alternatively a continuous diffusion process may be used to increase the total level of tannin and sugar extracted from the apple pulp. Residual pomace is frequently used for extraction of food-grade pectin; or it may be used as animal feed or for mulching.

Great care is needed to ensure that the starch content of the fruit has been convertd to sugar before milling, otherwise significant quantities of starch in the juice will affect both the economics and the practical aspects of fermentation and subsequent clarification. However, overripe fruit can

Figure 5.3 An 'apple canal' into which fruit is tipped either from bulk carriers, farm trailers or sacks. Above the canal is a conveyor carrying fruit from a bulk discharge area to the appropriate part of the canal.

give rise to significant problems during milling and pressing by modern equipment, especially in a continuous belt press.

Freshly pressed juice will normally be treated with *c*. 50 mg/kg sulphur dioxide to inhibit both oxidative browning and the growth of wild yeasts (qv). After a period of 24 hours it is normal to readjust the juice to give a low residual level (10–30 mg/kg) of free sulphur dioxide. Some juice will be evaporated to provide a juice concentrate that can be stored for fermentation at any time during the year rather than just at the harvest period.

5.2.3 Fermentation of cider

Fermentation vessels. Traditionally, the apple juice was fermented in oak barrels (Figure 5.7) or vats (Figure 5.8); although many such vats are still

Figure 5.4 Traditional pressing of cider apple pulp using a hydraulic screw press. The 'cheese' of apple pulp contained in a hessian cloth is held on a series of wooden slats.

in use, vats of concrete or mild steel with a ceramic or resin lining, glass-reinforced plastic and, more recently, stainless steel are used. Most cider vats are typically of 10 000–200 000 gallon capacity, but much larger vats exist. The largest fermentation/storage vat in the world is located at the premises of H P Bulmer Ltd in Hereford, (UK); this vat holds some 1.6 million gallons of cider. A few cider-makers use modern conico-cylindrical vats (Figure 5.9). However, tall narrow conico-cylindrical vats, such as those used by brewers, are not ideal. Jarvis, Forster and Kinsella (1995) demonstrated an effect of hydrostatic pressure on the fermentative ability of *Saccharomyces cerevisiae* in cider-making. At pressures in excess of 1.5 bar (equivalent to a fermenter height of 48 feet (15 metre)), the fermentation rate is slowed significantly and the flavour of the fermentation product also changes. The fermentation of cider is more akin to the fermentation of wine, and vats of limited height, such as the Unitank-style vessel, minimize the effect of hydrostatic pressure stress on the fermentation yeast.

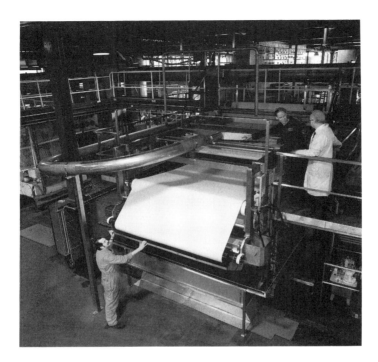

Figure 5.5 Installation of a modern continuous belt press at H P Bulmer Ltd, Hereford, UK.

Fermentation substrate. Traditional cider-making used whole juice, often with much of the apple solids remaining. Such solids included the pips, which contain cyanogenic glucosides, hence the cider could contain small quantities of cyanide derivatives. Folklore has it that farm cider-making was improved by the addition of dead rats or other animal corpses to the cider vat; such apocryphal stories abound but there is evidence that farmers would often suspend a leg of ham in a cider vat to 'enhance the curing process'. One of the consequences would be to increase the available nutrients and thereby stimulate the fermentation process.

In countries such as France where the permitted production processes are defined in law, only fresh apple juice or fresh juice reinforced with secondary juices can be fermented into cider. However, the process of chaptalization has become increasingly common in many countries where it is not proscribed (e.g. UK, Ireland, Belgium, Australia). Chaptalization is the process of supplementation of the basic juice with a suitable fermentation sugar (e.g. glucose syrup) which enables the production of a strong cider with an alcohol content up to 12% ABV. By comparison, fresh

Figure 5.6 Continuous pressing of apple pulp in a belt press.

Figure 5.7 Traditional cider making in oak barrels, Herefordshire, UK, 1935.

CIDER, PERRY, FRUIT WINES AND OTHER BEVERAGES 107

Figure 5.8 Traditional oak cider vats at H P Bulmer Ltd, Hereford, UK.

juice fermentations, depending solely on the fruit sugar content, will rarely exceed 6–7% ABV.

In the preparation of cider, the apple juice will be fresh or reconstituted from an apple juice concentrate; in many countries the apple juice can be blended with up to 25% by volume of pear juice (either perry pear (qv) or culinary pear juice); such blending produces a juice with distinctly different properties because of the high sorbitol content of pear juice. Other than fermentation sugars, the only other primary addition will be a suitable yeast culture. However, in fermentation brews originating wholly or largely from juice concentrates, it is normal to add yeast nutrients, such as diammonium phosphate, sodium pantothenate and thiamine.

The fermentation process. The fermentation of apple juice to cider can occur naturally through the metabolic activity of the yeasts and bacteria

Figure 5.9 A modern cider fermentation hall at H P Bulmer Ltd, Hereford: the photograph shows the base cones on a number of 150 000 gallon stainless steel fermentation vessels.

present on the fruit at harvest, which are then transferred into the apple juice on pressing. Other organisms, arising from the milling and pressing equipment and the general environment, can also contaminate the juice at this stage. Unless such organisms are inhibited, e.g. by maintaining 10–30 mg/l free sulphur dioxide, the mixed fermentation will yield a product which varies considerably from batch to batch, even if the composition of the apple juice is constant. Hence, control of the indigenous and adventitious microorganisms followed by deliberate inoculation with a selected strain of yeast is the preferred commercial route for the production of cider. However, there is a tendency for addition of an excess of sulphur dioxide to the juice such that many fermentation cultures are themselves partially inhibited, thus nullifying the effects of using selected culture strains. Indeed, in such circumstances, sulphur dioxide-resistant wild yeasts can then grow unchecked and produce undesirable flavour and aroma changes in the cider.

The process consists of partially filling the vat with juice followed by inoculation with a specific starter culture of yeast (frequently a wine yeast capable of growth at elevated alcohol levels). Once fermentation has started, additional substrate (apple juice, concentrate or fermentation sugar) is added from time to time until the vat is filled. The fermentation process typically continues without control of temperature, pH or other parameters until all the fermentable sugar has been metabolized into alcohol. Depending upon conditions, this process can take 2–12 weeks! There is no doubt that better control of temperature can give more consistent fermentation profiles and products. Whilst ambient temperature fermentation occurs widely, increasingly commercial cider-makers control temperature within a range of 20–25°C. In France and Belgium, cider is often fermented at a temperature of 15–18°C.

Secondary fermentation and maturation. When fermented to dryness, the cider is frequently left for a few days on the lees to permit the yeasts to autolyse, thereby adding cell constituents such as enzymes, amino acids and nucleic acids, to the brew. The cider will be racked (separated) from the lees and transferred either directly or after clarification into storage vats (traditionally made of oak). A secondary malo-lactic fermentation (qv) process occurs either in parallel with the primary yeast fermentation (typical of the process of cider-making in northern Spain) or following completion of the primary fermentation. In addition, storage of the cider following fermentation results in a number of chemical changes that affect significantly the flavour of the cider. Judgement as to the extent of maturation and the suitability of the cider for use is still an art vested in the craft skill of the cider-maker.

Final processing. When required for use, different batches of cider, generally made from mixtures of different juices, will be blended by the cider-maker to provide specific aroma and flavour attributes. Traditionally, the raw cider will be fined using agents such as bentonite, gelatin or chitin and filtered to give a bright product with no haze. Modern processing refinements include the use of cross-flow or other microfiltration systems to obviate the need for fining and to speed the process.

If the cider has a high alcohol content (e.g. 9–12% ABV) it may be 'broken back' to the final product strength using water or dilute apple juice. Other ingredients, such as sugars, intense sweeteners (e.g. saccharin), colours and/or additional preservative (in the EU this is restricted to sulphur dioxide or sorbic acid) may be added (Anon, 1994a,b; 1995). The finished blend may be carbonated or subjected to further processing prior to packaging.

Cider quality and legislation. Legislative criteria in some countries (e.g. France and Spain) define not only the raw materials and processes that may be used in cider-making but also lay down mininum analytical criteria for the products. In the UK, the only specific legislation controlling cider-making (other than the general requirements of food legislation) are the Official Notices produced by the Commissioners for Excise. In the absence of official quality criteria, the National Association of Cider Makers

(NACM) has operated a voluntary Code of Practice for British Cider and Perry for many years. The code defines, *inter alia*, approved ingredients and processes for cider (and perry) and lays down minimum quality criteria with which members of the NACM must comply. A regular monitoring exercise is undertaken by the NACM to ensure compliance with the Code of Practice, which requires that products contain:

actual alcohol	>1.2 and <8.5 g/100 ml (ABV)
sugar-free dry extract	not less than 13 g/l
volatile acidity (as acetic acid)	not more than 1.4 g/l
iron	not more than 7 mg/l
copper	not more than 2 mg/l
arsenic	not more than 0.2 mg/l
lead	not more than 0.2 mg/l

These criteria are not dissimilar to those laid down by legislation in other countries.

5.2.4 Special types of cider

Low-alcohol cider (<1.2% ABV). Low-alcohol ciders have been developed either by dealcoholization of strong cider or by using a 'stopped fermentation'. Dealcoholization uses processes such as falling film evaporation or reverse osmosis. The dealcoholized base lacks flavour and body and is unsuitable for sale without considerable improvement. One such product, which has been marketed for several years, uses the dealcoholized cider as a base which has then been fortified with cider, apple juice flavours and selected alcohols to rebuild both the flavour and body of the product. The 'stopped fermentation' method (*c.* 2% ABV) racks the cider off and chills and clarifies it before completion of the process. The cider is then diluted with fresh unfermented apple juice to give a product having some cider flavour.

'White' cider (generally 6.5–8.4% ABV). White ciders are prepared by either fermentation of charcoal-decolorized apple juice or by decolorization of the fermented cider. Decolorization of apple juice also removes many juice flavours and flavour precursors such that the fermented product has a softer, less cidery flavour than does a product made by decolorization of fermented cider. By the careful selection of decolorizing agents (e.g. PVPP), other than charcoal, it is possible to minimize the loss of flavour compounds.

'Black' cider. Black cider was introduced to the UK market on a trial basis in 1992; it consists of a co-fermentation of apple juice with hops and a

heavily malted barley. The flavour is very complex, having overtones of both cider and stout. Although not widely developed, it demonstrates the flexibility of cider as a base material for the development of a range of new alcoholic beverages.

'Ice' cider. Following the successful development of various processes for the production of 'ice beers', the concept has been extended to cider. The details of the processes used are currently confidential, but cooling of cider to a temperature at which ice crystals form and the subsequent separation of the cider from the ice results in subtle flavour changes. However, the chemical nature of the changes is not yet fully understood.

'Vintage' and 'single cultivar' ciders. A cider fermented from the fresh juice of fruit harvested in a named year can be designated as a vintage cider. Similarly, cider prepared from a single apple cultivar, e.g. the highly astringent Kingston Black apple, can be designated to be a Single Cultivar Cider. Combination of the two can thus generate a Single Cultivar Vintage Cider.

Cidre nouveau. A recent marketing development in both England and France has been the introduction in early November of a freshly fermented (i.e. non-matured) cider given the designation 'cidre nouveau'. Such products are available for sale for a few months only (in France until the following March).

Cidre bouché and cidre fermette. Traditional French farmhouse cider, sold in bottles with a wired cork and prepared using only traditional ingredients and processes, is sold in France as cidre bouché. Generally, such products have an alcohol content of *c.* 2% ABV. A similar product is manufactured in Belgium, where it is known as cidre fermette (farmhouse cider).

5.2.5 *The microbiology of apple juice and cider*

Freshly pressed apple juice contains a variety of yeasts and bacteria, many of which will be incapable of growth at the acidity of the juice. Examples of organisms often present in juice are shown in Table 5.3, together with an indication of their susceptibility to sulphur dioxide.

The role of sulphur dioxide in apple juice. The use of sulphur dioxide as a preservative in cider-making is controlled by legislation; in the EC the maximum level permitted in the final product is 200 mg/l (Anon, 1995). The addition of sulphur dioxide to apple juice results in the formation of so-called sulphite addition compounds through the binding of sulphur

Table 5.3 Typical microorganisms of fresh apple juice

Type	Typical species	Sensitivity to sulphur dioxide
Yeast	*Saccharomyces cerevisiae*	± or −
	Saccharomyces uvarum	± or −
	Saccharomycodes ludwigii	−
	Kloeckera apiculata	+++
	Candida pulcheriima	++++
	Pichia spp.	++++
	Torulopsis famata	++
	Aureobasidium pullulans	+++
	Rhodotorula spp.	++++
Bacteria	*Acetobacter xylinum*	++
	Pseudomonas spp.	++++
	Escherichia coli	++++
	Salmonella spp.	++++
	Micrococcus spp.	++++
	Staphylococcus spp.	++++
	Bacillus spp.	− (spores)
	Clostridium spp.	− (spores)

Sensitivity: −, insensitive; +, relatively insensitive; ++, +++, ++++ increasingly more sensitive.
From Jarvis (1993b).

dioxide to carbonyl compounds. When dissolved in water, sulphur dioxide and its salts set up a pH-dependent equilibrium mixture of 'molecular sulphur dioxide', bisulphite and sulphite ions (Beech and Carr, 1977; Beech and Davenport, 1983). The antimicrobial activity of sulphur dioxide is believed to be caused by the molecular sulphur dioxide moiety of the part that remains unbound (the so-called 'free' sulphur dioxide). Less sulphur dioxide is needed in juices of high acidity, for instance 15 mg/l of free sulphur dioxide at pH 3.0 has the same antimicrobial effect as 150 mg/l at pH 4.0.

The binding of sulphur dioxide is dependent upon the nature and origin of the carbonyl compounds present in the juice. Naturally occurring compounds that bind sulphur dioxide include glucose, xylose and xylosone. If the fruit has undergone some degree of rotting, other binding compounds will be present, including 2,5-dioxogluconic acid and 5-oxofructose (2.5-D-*threo*-hexodiulose). Such juices will require increased additions of sulphur dioxide if wild yeasts and other microorganisms are to be controlled effectively.

The addition of sulphur dioxide to a fermenting juice results in rapid combination with acetaldehyde, pyruvate and α-oxoglutarate, produced by the fermenting yeasts. Consequently, all such additions must be completed immediately after pressing the juice although, provided the initial fermentation is inhibited, further additions to give the desired level of free sulphur dioxide to inhibit wild yeasts can be made during the following 24 hours.

Fermentation yeasts. The fermentation process is carried out by strains of *Saccharomyces* spp., especially *S. cerevisiae*, *S. bayanus* and *S. uvarum*, which are added to the sulphited juice as a pure culture. Traditionally, the starter culture is prepared in the laboratory, although, increasingly, commercially produced dried yeast cell preparations are used, either for direct vat inoculation or as inocula for the yeast propagation plant.

The choice of culture is dependent upon many criteria, such as flocculation characteristics, ability to ferment efficiently at sub-ambient temperatures, alcohol and sulphur dioxide tolerance and lack of ability to produce hydrogen sulphide (Beech, 1993). One desirable characteristic is the ability to produce fusel oils (e.g. higher alcohols), which affect both the flavour and aroma of the cider (Table 5.4).

The fermentation process typically takes some 2 to 12 weeks to proceed to dryness (i.e. to a specific gravity of 0.990–1.000), at which time all fermentable sugars have been converted to alcohol, CO_2 and other metabolites. After inoculation, the starter yeast together with those sulphur dioxide-resistant wild yeasts selected from the juice will increase in numbers from an initial level of about $(0.2–1) \times 10^6$ cfu/ml to $(2–5) \times 10^7$ cfu/ml. Following an initial aerobic growth phase, the resulting oxygen limitation and high carbohydrate levels in the media trigger the onset of the anaerobic fermentation process.

In controlled fermentations, a maximum temperature of 25°C will generally be permitted, although fermentations controlled at 15–18°C are not uncommon, especially in France where long slow fermentations are often preferred. Temperatures higher than 25°C are generally undesirable, since metabolism by the desired yeast strain may be inhibited, leading to 'stuck' fermentations and the growth of undesirable thermoduric yeasts and spoilage bacteria. 'Stuck' fermentations can sometimes be restarted by addition of nitrogen (10–50 mg/l), usually as ammonium sulphate or diammonium phosphate, sterols (usually in the form of yeast hulls) and/or thiamine (0.1–0.2 mg/l).

Table 5.4 Higher alcohols in apple juices and ciders

Constituent	Concentration range (mg/l)	
	Apple juices	Ciders
Propan-1-ol	0.2–9	4–200
Butan-1-ol	0.5–24	4–32
Butan-2-ol	<0.1	14–74
Heptan-1-ol	0.1	42–196
Octan-1-ol	0.1–2	16–39
Hexan-1-ol	1–2	2–17
2-Phenylethanol	<0.1	7–260

Modified from Jarvis (1993b).

Maturation and secondary fermentation. The maturation vats are filled with the racked-off cider and either provided with an 'over-blanket' of CO_2 or otherwise sealed to prevent ingress of air, which would stimulate the growth of film-forming yeasts (e.g. *Brettanomyces* spp., *Pichia membranefaciens*, *Candida mycoderma*) and aerobic bacteria (e.g. *Acetobacter xylinum*). Growth of the former will produce precursors for the development of defects such as 'mousey' flavour (e.g. 1,4,5,6-tetrahydro-2-acetopyridine) whilst the latter will acetify the cider through the production of acetic and other volatile acids, which imparts a vinegary note to the product. Of course, deliberate acetification of cider can be used to produce cider vinegar.

During the maturation process, growth of lactic acid bacteria (*Lactobacillus pastorianus* var. *quinicus*, *L. mali*, *L. plantarum*, *Leuconostoc mesenteroides*, *Leuc. oenus*, etc.) can occur extensively, especially in wooden vats. The control of the malo-lactic fermentation processes in cider is only now beginning to be understood (see Jarvis *et al.*, 1995). The 'malo-lactic fermentation' process results in the conversion of malic and quinic acids to lactic and dihydroshikimic acids, respectively. Such secondary fermentation reduces the acidity of the cider and imparts subtle changes that are generally considered to improve the flavour of the product. However, in certain circumstances, some metabolites of the lactic acid bacteria may damage the flavour and result in spoilage, e.g. excessive production of diacetyl (and its vicinal-diketone precursors) (Table 5.5), the 'butterscotch-like' taste of which can be detected in cider at a threshold level of about 0.4 mg/l.

Swaffield, Scott and Jarvis (1994) studied the microbial colonization of cubes of sterilized old vat oak suspended in a vat of maturing cider. Over a period of several weeks it was found that organisms are able to penetrate into and colonize to a depth of at least 1 cm from the surface of the timber. The organisms were diverse strains of both yeasts and bacteria, including many that exude mucilaginous materials. This is of some significance in relation to the manner in which entrapped microorganisms cause changes in cider during maturation; it suggests that the secondary maturation process in wooden vats may be a natural immobilized microbial fermentation. This opens up possibilities for introducing controlled immobilized fermentation of cider in the future.

Spoilage and other microorganisms in cider. Bacterial pathogens such as *Salmonella* spp., *Escherichia coli* and *Staphylococcus aureus* may occasionally occur in apple juice, where they will have been derived from the orchard soil, farm and process equipment or human sources. However, the acidity of the product prevents growth and such organisms do not survive for long in the fermenting product. Bacterial spores from species of *Bacillus* and *Clostridium* can survive for long periods and are frequently

Table 5.5 Volatiles in normal and diacetyl-spoiled cider[a]

	Compound	Normalised peak area	
		Normal	Spoiled
Alcohols	2-Methyl propanol	35.4	97.3
	Iso-amyl alcohol	305	213
	2- and 3-Methyl-butan-1-ol	503	456
	Heptan-1-ol	1.26	0.75
	Hexan-1-ol	35.7	29.5
	Nonan-1-ol	0.79	0.86
	Methionol	1.29	0.55
	2-Phenylethanol	69.0	61.6
Esters	Ethyl acetate	86.1	89.2
	Ethyl lactate	45.9	37.9
	Ethyl-2-methylbutyrate	10.3	12.9
	Ethyl-2-hydroxy-4-methyl pentanoate	4.19	7.03
	Ethyl benzoate	4.70	5.66
	Ethyl hexanoate	233	179
	Ethyl octanoate	280	226
	Ethyl decanoate	57.6	53.4
	Ethyl dodecanoate	2.10	1.04
	Diethyl succinate	5.23	2.88
	Hexyl acetate	1.95	6.25
	Hexyl octanoate	3.39	11.5
	2-Phenylethyl acetate	11.4	6.42
Aldehydes	Benzaldehyde	1.42	1.41
	Octanal	0.67	1.19
	Decenal	7.52	0.36
	Undecanal	2.16	1.23
Ketones	Decan-2-one	3.58	0.90
Acids	Hexanoic acid	12.6	12.6
	Heptanoic acid	0.00	3.39
	Octanoic acid	34.6	29.4
	Nonanoic acid	1.01	1.21
Miscellaneous	Diacetyl	0.00	3.42
	γ-Decalactone	4.39	3.39
	Ethyl guaiacol	2.47	4.53
	Eugenol + 4-ethylphenol	1.86	1.52

[a]Based on GC–MS analysis of headspace volatiles.
Modified from Jarvis (1993b).

found in cider but are of no concern in relation to either spoilage or public health because of the low pH value of cider; nonetheless their presence may be indicative of poor plant hygiene.

A recent report from the USA of a serious outbreak of cryptosporidiosis following the consumption of freshly pressed apple juice (*sic* cider in the USA) does not relate to contamination of alcoholic cider by oocysts of the intestinal parasite *Cryptosporidium*. Infection by this newly emerging pathogen arose from faecal contamination of apples collected from the ground in an orchard in which cows had recently grazed (Millard *et al.*,

1994). Unpublished studies in the UK have shown that oospores of *Cryptosporidium* are sensitive to pasteurization and to sulphur dioxide; in addition, filtration processes, such as those used in the commercial production of cider, would remove any ooscysts if they did occur in the juice.

The juice from unsound fruits and from contamination within the pressing plant may show extensive contamination by microfungi, such as *Pencillium expansum*, P. crustosum, Aspergillus niger, *A. nidulans*, *A. fumigatus*, *Paecilomyces varioti*, *Byssochlamys fulva*, *Monascus ruber*, *Phialophora mustea* and by species of *Alternaria*, *Cladosporium*, *Botrytis*, *Oospora* and *Fusarium*. None is of particular concern in cider-making, except that thermo-resistant spores of organisms such as *Byssochlamys* spp. can survive pasteurization and may grow in cider if it is not adequately carbonated.

The occurrence of the mycotoxin patulin in mould-rotted apples, following the growth of *Penicillium expansum* and other organisms, may lead to carry over of patulin into the cider base, but patulin is destroyed during the fermentation process. Investigations by Moss and Frank (unpublished data) indicate that the ability of yeast to metabolize patulin is inducible and that possible breakdown compounds include æscladiol and patulin methyl ester.

Reference has been made to the role of organisms such as *Brettanomyces* spp. and *Acetobacter xylinum* in the spoilage of ciders during the later stages of fermentation and maturation. Of equal concern is the yeast *Saccharomycodes ludwigii*, which is often resistant to sulphur dioxide levels as high as 1000–1500 mg/l. Maloney (1993) showed that isolates of the yeast *S. ludwigii* were very tolerant of sulphur dioxide because of the production of an excess of acetaldehyde, which was able to minimize the accumulation of intracellular sulphite. He also found that sulphur dioxide-resistant strains had relatively low intracellular pH values and a low intracellular buffering capacity.

S. ludwigii can grow slowly during all stages of fermentation and maturation and is often an indigenous contaminant of cider-making premises. Its presence in bulk stocks of cider does not cause overt problems. However, if it is able to contaminate 'bright' cider at bottling, its growth will result in a butyric flavour and the presence of flaky particles that spoil the appearance of the product. Although single cells of the organism are sensitive to pasteurization conditions, clumps of organisms can survive pasteurization, thereby contaminating the final product at the packaging stage; contamination can also arise from the packaging plant and the environment.

Environmental contamination of final products can also occur from other yeasts such as *S. cerevisiae*, *S. bailii* and *S. uvarum*, which will metabolise any residual or added sugar to generate further alcohol and, more importantly, to increase the concentration of CO_2. Strains of such environmental contaminants are frequently resistant to sulphur dioxide. Bottles of cider inoculated with such fermentative organisms may develop carbonation pressures up to 9 bar. For this reason, it is essential to minimize any risk from contamination of packaged product by fermentative yeasts and to maintain an adequate level of free sulphur dioxide in the final product. This is particularly important in multiserve containers, which may be opened and then stored with a reduced volume of cider. This precaution is less important for product packaged in single-serve cans and small bottles, which receive a terminal pasteurization process after filling.

Special secondary processes. Traditional 'conditioned' draught cider results from a live secondary fermentation process. After filling into barrels, a small quantity of fermentable carbohydrate is added to the cider followed by an inoculum of active alcohol-resistant yeasts. The subsequent growth is accompanied by a low-level fermentation during which sufficient CO_2 is generated to produce a *pettilance* in the cider together with a haze of yeast cells. Such products have a relatively short shelflife in the barrel.

Double fermented cider is initially fermented to a lower than normal alcohol content (e.g. 6% ABV) by restricting the amount of chaptalization sugar added. The liquor is racked off as soon as the cider has fermented to dryness and is either sterile-filtered or pasteurized prior to transfer to a second sterile fermentation vat. Additional sugar or apple juice is added and a secondary fermentation is induced by inoculation with a different yeast strain, e.g. an alcohol-tolerant strain of *Saccharomyces* spp. Such a process permits the development of very complex flavours in the cider.

Sparkling ciders are normally prepared nowadays by artificial carbonation to a level of 3.5–4 vol. CO_2. Traditionally, sparkling ciders were prepared according to the 'méthode champenoise'. After bright filtration of the fully fermented dry cider, it would be filled into bottles containing a small amount of sugar and an appropriate champagne yeast culture. The bottles were corked and wired and laid on their side for the fermentation process, which lasts 1–2 months at 15–18°C. Following this stage, the bottles were placed in special racks with the neck in a downwards position. The bottles were gently shaken each day to move the deposit down towards the cork, a process that can take up to 2 months. The disgorging process involved careful removal of the cork and yeast floc but without loss of any liquid (sometimes the neck of the bottle would be frozen to aid this process). The disgorged product would then be topped up using a syrup of alcohol, cider and sugar prior to final corking, wiring and labelling. It is not difficult to understand why this process is rarely used nowadays! An alternative procedure, which gives a gentle 'pettilance' similar to that of the traditional 'méthode champenoise' is to undertake a secondary 'cuvée close' fermentation in a sealed pressurized vat so that the CO_2 generated is retained within the cider.

5.2.6 The chemistry of cider

The chemical composition of cider is dependent upon the composition of the apple juice, the nature of the fermentation yeasts, microbial contaminants and their metabolites, and on the nature of any additives used in the final product.

The composition of cider apple juice. Apple juice is a mixture of sugars (primarily fructose, glucose and sucrose), oligosaccharides and polysacharides (e.g. starch) together with malic, quinic and citromalic acids, tannin (i.e. polyphenols), amides and other nitrogenous compounds, soluble pectin, vitamin C, minerals and a diverse range of alcohols, aldehydes, ketones, esters, fatty acids, and hydrocarbons. Some esters (e.g. ethyl 2-methylbutanoate) give the juice a typical apple-like aroma (Paillard, 1990). The relative proportions are dependent upon the variety of apple, the cultural conditions under which it was grown, the state of maturity of the fruit at the time of pressing, the extent of physical and biological damage (i.e. mould rots) and, to a lesser extent, the efficiency with which the juice was pressed from the fruit.

Treatment of the fresh juice with sulphur dioxide results in the complexing of carbonyl compounds to form stable hydroxysulphonic acids. If the apples contained a high proportion of mould rots then appreciable amounts of carbonyls such as 2,5-dioxogluconic acid and 2,5-D-*threo*-hexodiulose will occur. Sulphur dioxide is important also in the prevention of enzymic and non-enzymic browning reactions of the polyphenols.

Products of the fermentation process. The primary objective of fermentation is the production of ethyl alcohol from simple sugars. Various intermediates in the Embden–Meyerhof–Parnass pathway can also be converted to form a diverse range of other metabolites, including glycerol (up to 0.5%). Diacetyl and acetaldehyde may also occur, particularly if the process is inhibited by excess sulphite and/or if uncontrolled malo-lactic fermentation occurs. Other metabollic pathways will operate simultaneously with the formation of long- and short-chain fatty acids, esters, lactones, etc. Methanol will be produced in small quantities (10–100 mg/l) as a result of demethylation of pectin in the juice. Table 5.5 illustrates some of the volatile compounds found in a normal and a spoiled cider.

If other organisms are also present in the fermentation (e.g. lactic acid bacteria), these can convert the fruit acids malic and quinic acids to lactic and dihydroshikimic acids, respectively, thereby reducing the acidity of the cider by about 50%. These reactions are accompanied by further diverse but not widely understood chemical changes that result in subtle, yet important, flavour changes in the final product. Lactic and acetic acids can also be formed directly from residual sugars and great care needs to be taken to avoid excessive production of volatile acids in cider.

The tannins in cider do not change significantly during fermentation, although the chlorogenic, caffeic and *p*-coumaryl quinic acids may be reduced to dihydroshikimic acid and ethyl catechol.

The nitrogen content of cider juice comprises a range of amino acids, the most important of which are asparagine, aspartic acid, glutamine and glutamic acid; smaller amounts of proline and 4-hydroxymethylproline also occur. Aromatic amino acids are virtually absent from apple juices. With the exception of proline and 4-hydroxymethylproline, the amino acids will be largely assimilated by the yeasts during fermenation. However, leaving the cider on the lees for an appreciable length of time will significantly increase the amino nitrogen content as a consequence of release of cell constituents during autolysis.

Inorganic compounds in cider are derived largely from the fruit and will depend upon the conditions prevailing in the orchard. These levels will not change significantly during fermentation. Small amounts of iron and copper may also occur naturally but the presence of more than a few milligrams per litre will result in significant black or green discolorations and flavour deterioration. Such discolorations are caused by the formation of iron or copper tannates from traces of metal ions derived from equipment and/or from use of rotten fruit.

Cider maturation and cider flavour. Changes in the composition of the cider occur during maturation, but the nature of the changes is but poorly understood. The flavour of cider is very complex and is dependent upon the source, type and pretreatment of the juice, the nature of the yeast strain(s) used, the type and conditions of the fermentation processes (including the occurrence or otherwise of the malo-lactic fermentation) and many other variables. The primary attributes for assessing cider flavour are related to descriptive notes concerned primarily with astringency, bitterness, fruitiness, sweetness, sourness and acidity; other related attributes include the bouquet, the alcohol content and sulphur notes. Off-flavours include those described as 'mousey', musty, yeasty, diacetylic, sulphury, 'rotting vegetable', acetic and metallic (oxidized).

Unlike wines, and spirits such as whisky, the wooden vats traditionally used for maturation contribute few, if any, flavour compounds to the cider; however, they are an important source of organisms for the secondary fermentation and maturation processes, which can produce a diversity of flavour characteristics. Analysis of green, fermented and matured ciders has so far led to the identification of over 200 volatile and non-volatile compounds that individually and in combination affect the flavour of the product. Some of the factors affecting the occurrence of cider flavour compounds have been reported recently (Jarvis *et al.*, 1995) and this is an area of on-going investigation.

5.3 Perry (poire)

The production of perry parallels that of cider in most respects. Perry production is traditional in certain parts of the UK (especially Gloucestershire and Somerset) and in France (Normandy). The fermentation base can be blended with up to 25% of apple juice.

As with the production of cider, special varieties of perry pears are grown traditionally (Table 5.6) especially in Gloucestershire. Perry pears have a low to moderate content of tannin and introduce a high degree of astringency to the perry that is not found in fermented culinary pear juice. Although fewer studies have been made on pears than on apples, it has been suggested by Paillard (1990) that pear aromas contain significantly fewer volatile constituents (79) than does the aroma of apple (212). Pear juice contains a higher proportion of acetate and other short-chain aliphatic esters than does apple juice, with fewer alcohols, higher esters or lactones. However, in many varieties of pear, the aroma is characterized by a range of deca-2,4-dienoate esters, all of which possess a powerful pear aroma. The methyl- and ethyl-deca-2,4-dienoates have high boiling points and are retained even in concentrated pear juices.

Because it was feared that the genetic base of traditional perry pears would be lost, a national collection of perry pears varieties has been established recently at the Three Counties Agricultural Showground at Malvern in Worcestershire, UK.

Table 5.6 Characteristics of some perry pear varieties

Type	Typical varieties	Typical composition	
		Acidity (g/100 ml)	Tannin (g/100 ml)
Low acid, low tannin	Red Pear	0.29	0.09
Low acid, medium tannin	Late Treacle	0.33	0.15
	Nailer	0.37	0.23
Medium acid, low tannin	Brown Bess	0.55	0.10
	Gregg's Pit	0.57	0.11
Medium acid, medium tannin	Blakeney Red	0.42	0.13
	Moorcroft	0.50	0.17
	Tumper	0.53	0.15
High acid, low tannin	Yellow Huffcap	0.62	0.10
High acid, medium tannin	Holmer	0.84	0.30

Data from Williams, Faulkner and Burroughs (1963).

Details of the production of perry are given by Pollard and Beech (1963). The major difference between perry and cider is in the flavour: perry has a much 'softer' and less astringent character and the occurrence of a high level of non-fermentable sorbitol gives the product increased body. Perry stored for a long time will tend to 'harden' and lose the smooth flavour characteristics; the reasons for such changes are unknown but they are probably related to instability of the polyphenols.

5.4 Fruit wines

The production of fruit wines is undertaken commercially in Europe and in many other parts of the world, especially those unsuited to the cultivation of grape vines. In Europe, the Association of Cider and Fruit Wines Producers of the EEC has agreed definitions, for Excise purposes, of two types of fruit wine: fruit wines produced with and without fortification with alcohol.

Fruit wine without addition of alcohol is defined as:

> An alcoholic beverage obtained by the complete or partial fermentation of the fresh, concentrated or reconstituted juice or pulp of fresh edible fruits (domestic or tropical), other than grapes, with the addition of water, sugar or honey. Fresh, concentrated or reconstituted juice may be added also after fermentation. Fruit wines will have an alcoholic strength in the range between 8 and 14% ABV. Fruit wines may be uncarbonated (i.e. still) or carbonated by the injection of carbon dioxide or by a secondary fermentation.

Fruit wine with added alcohol is defined as:

> An alcoholic beverage obtained by the complete or partial fermentation of the fresh, concentrated or reconstituted juice or pulp of fresh edible fruits (domestic or tropical) or other parts of fresh plants, other than grapes, with or without the addition of water, sugar and agricultural alcohol. Fresh, concentrated or reconstituted juice, or flavours, may be added after fermentation. Fruit wines will have an alcoholic strength in the range between 8 and 20% ABV; most will have an actual alcohol content in the range 12 to 15% ABV. Fruit wines may be uncarbonated (i.e. still) or carbonated by the injection of carbon dioxide or by a secondary fermentation.

Fruit wines produced in accordance with these definitions will be coloured and/or flavoured by the juices used in their manufacture; they may also be further coloured by the addition of permitted food colours (Anon, 1994b). By contrast, fruit wines manufactured in Eastern European countries such as Poland will not necessarily be perceived as being different from grape wines, being generally white, rosé or red in colour and classified according to their alcohol and sugar contents. Jarczyk and

Table 5.7 Proximate composition of some fruit wines

Style of wine	Country of origin	Alcohol (g/100 ml)	Sugar (g/100 ml)	Acidity (g/l)	SFDE[a] (g/l)
Dry	Former Czechoslovakia	10.0	<2.0	ND	ND
	Poland	12.5	6.6	5.8	ND
	Former USSR	12.5	2.5	8.0	22.0–30.1
Medium sweet	Former Czechoslovakia	12.0	3.8	ND	ND
	Poland	13.0	10.4	6.6	ND
	Former USSR	13.0	6.3	7.8	25.0–31.5
Sweet	Former Czechoslovakia	14.0	10.0	ND	ND
	Poland	13.8	13.3	6.9	ND
	Former USSR	13.5	10.2	8.2	30.6–35.1

ND: no data.
[a]SFDE: sugar-free dry extract.
Data from Jarczyk and Wzorek (1977).

Wzorek (1977) noted that legislative definitions exist in Poland, the former Czechoslovakia and the former USSR for different styles of fruit wine. Some examples of fruit wine compositional data are presented in Table 5.7.

As with cider, compositional standards relating to maximum volatile acidity and sugar-free dry extract are prescribed to prevent excessive dilution of the products. For example, the maximum permitted level of volatile acidity (measured as acetic acid) should not exceed 1.2 g/l for white and 1.6 g/l for red wines (Jarczyk and Wzorek, 1977).

5.4.1 Fruits used in fruit wine manufacture

In Western and Central Europe, fruit wines are prepared by the fermentation of cultivated fruits, such as apples, cherries, currants, pears, plums and strawberries, and wild fruits, such as bilberries, blackberries, elderberries and rose hips. Wild fruits frequently contain higher levels of acidity and tannins; they also frequently have stronger aromas than cultivated varieties of the same plants. Similar fruits have been used in other countries also. In recent years, considerable attention has been paid to the use of sub-tropical and tropical fruits for the production of wines: these include apricots (Joshi *et al.*, 1990), banana (Mosso *et al.*, 1990), carambola (Poullet, 1994), kiwi (Broussous and Ferrari, 1990; Tjomb, 1991), mango (di Baggi *et al.*, 1986; Adesina, Oguntimein and Obisanya, 1993), muskmelon (Teotia *et al.*, 1991), orange (Rose, 1991), papaw and plantain (Mosso *et al.*, 1990), and persimmon (Gorinstein *et al.*, 1993).

Compositional data for a variety of fruits used in wine-making are summarized in Table 5.8.

Table 5.8 Proximate compositional analysis of edible parts of fruits used in winemaking

Fruit	Total sugar (g/100 ml)	Total acidity (g/l)	Tannin (g/100 ml)
Apple (*Pirus malus*)	10.0–11.2	6.0–7.0[a]	0.1–0.50
Blackberry (*Rubus fructicosus*)	5.5–12.8	9.0–12.0[b]	0.3
Cherry, sweet (*Prunus avium*)	9.5–16.6	6.0–12.0[c]	ND
Cherry, sour (*Prunus cerasus*)	10.0–12.2	13.0–18.0[c]	0.14
Black currant (*Ribes nigrum*)	7.0–15.4	10.0–21.0[b]	0.39
Plum (*Prunus domestica*)	9.3–12.0	9.0–12.0[a]	0.07
Raspberry (*Rubus idaeus*)	4.7–11.6	13.5–16.0[b]	0.26
Apricot (*Armeniaca vulgaris*)	6.7–8.5	7.0–13.0[a]	0.07
Peach (*Persica vulgaris*)	7.8–12.8	8.0–9.0[b]	0.10
Pear (*Pirus communis*)	9.5–13.0	3.0–5.0[a]	0.03–0.44
Mango (*Mangifera indica*)	11.3–15.4	2.4–6.8[a]	ND

ND, not determined.
Acidity expressed as: [a]malic acid; [b]citric acid; [c]tartaric acid.
Data from Nagy, Chen and Shaw (1993); Jarczyk and Wzorek (1977).

5.4.2 Processing of the fruit

Fruits for wine-making are taken to the winery where they are sorted, washed and macerated, or milled, prior to pressing. Components that will damage the flavour and aroma (e.g. stalks) are removed prior to maceration. After maceration, the fruit pulp will be pressed either in a rack-and-cloth hydraulic press (cf. cider-making) or in a modified Bücher–Guyer type press. Continuous belt and screw presses are used less extensively. The macerated fruit will be treated with a suitable enzyme preparation to break down the pectin and thereby increase the juice yield. In some instances, the pulp will be heated to 80–85°C during treatment with pectolytic enzymes prior to pressing. Such elevated temperatures will increase the colour of the juice, stabilize the anthocyanin pigments and destroy wild yeasts and other microorganisms (Rommel, Wrolstad and Heatherbell, 1992; Withy, Heatherbell and Fisher, 1993).

Juice not required immediately for pressing may be stored following the addition of sulphur dioxide at levels of up to 1200 mg/l, usually in the form of gaseous sulphur dioxide or as potassium metabisulphite. Alternatively, and preferably, the juice will be concentrated some seven-fold following depectinization and clarification; the concentrate can then be stored at chill temperature (7–10°C) until required. Berry fruits can also be deep frozen and stored at −18°C until required for pressing.

5.4.3 Fermentation of fruit wines

The fermentation process. As in the fermentation of cider, fruit wine fermentation is done largely in vats made of concrete, steel or GRP;

alternatively, traditional oak or larch wood casks may be used. The vat is partially filled with the prepared juice (desulphited if made from sulphur dioxide-preserved juice) to which will have been added part of any required fermentation sugar syrup (usually sucrose). Often the acidity of the juice will have been adjusted prior to use, either by the addition of food-grade acid or, more often, by the neutralization of excessive acidity using calcium carbonate. This is particularly important in the case of fruits or other plant extracts containing oxalic acid, which would otherwise cause difficulties in clarification of the wine at a later stage.

Since most fruit juices are relatively deficient in nitrogenous components, nutrients such as diammonium phosphate, ammonium chloride, ammonium sulphate or ammonium carbonate are added at a level of 0.1–0.3 g/l juice. Yang and Wiegand (1949) recommend the addition of urea at a level of 0.5 g/l to juices of berry fruits, such as raspberry and blackberry, to increase the fermentation rate. Yeast hydrolysate may also be added to provide a source of vitamins and sterols. Fermentation bases prepared from fresh, i.e. non-sulphited juice and non-concentrated juice, should be treated with some 30–150 mg/l sulphur dioxide to prevent growth of wild yeast strains.

The yeast inoculum will have been prepared in a pasteurized fruit juice base containing approximately 15% w/v sugar at 22–25°C. The prepared yeast culture will be added to the juice base to give $c.\ 10^6$ viable cells/ml in a volume equivalent to 1–10% of the total volume and at a yeast dry matter of 12.5 g/l. Normally the juice will have been prepared at a temperature of 12–15°C (never in excess of 20°C) prior to inoculation in order that the initial exothermic aerobic growth of the inoculum does not lead to excessively high temperatures with consequential auto-sterilization of the base. Over the first few days, the temperature will rise gradually to a level of 20–25°C and must, in any case, not be allowed to exceed 28°C. Additional sugar syrup will be added on two or three further occasions prior to the fermentation going to 'dryness' i.e. with no fermentable carbohydrate remaining. Dependent upon the level of alcohol produced and the fermentation conditions, the process may continue for some 6–8 weeks.

In some processes, fermentation may be carried out at a temperature of 5–10°C using a psychrophilic yeast strain (this is not dissimilar to the conditions used in the fermentation of champagne). Understandably, such a process takes much longer to ferment to dryness but preserves the fruit aroma rather better than does fermentation at higher temperatures. Cold fermentation also enables generation of a higher alcohol content and slows the rate of yeast autolysis.

Maturation and mellowing. At the end of fermentation, the wine is racked off the lees to remove both fruit solids and autolysing yeasts. The young wine will normally not have developed a definitive bouquet and requires maturation before packaging. Maturation of fruit wines is normally undertaken at a temperature of 7–15°C; further racking will be carried out after periods of one and two months and then at two-monthly intervals over a period of c. one year. If it is intended to produce very high-quality products, the wine would normally be racked every six months until the product is deemed to be suitably mature. During the maturation process, a secondary malo-lactic fermentation will normally occur, but only at the higher maturation temperatures. This secondary fermentation converts malic and citric acids into lactic and citromalic acids, respectively.

Subsequent mellowing of the wine results from the chemical and enzymic changes that take place throughout the maturation process and provides a diversity of esters, alcohols, aldehydes, ketones and acetals, which combine to provide the bouquet and flavour of the wine. Important in this process is the small amount of O_2 incorporated into the wine each time that racking is done and the further diffusion of O_2 through the staves of the barrel.

Final processing. The final stage of the process is the blending, sweetening and flavouring (if required) and stabilization of the wines. The blending process is done both to ensure consistency of product character and to reduce the strong aroma and flavour characters of certain wines. Although there is some preference for single-fruit wines, many are blends, especially with apple wine, which is relatively low in flavour.

Wines can be sweetened using either sugar or fruit juice, the latter also serving to increase the 'natural' fruit content and flavour of the wine. In some cases, it may be necessary also to adjust the acidity of the wine by the addition of an approved food-grade acid, such as citric or tartaric acid.

Clarification of wines prior to bottling involves the treatment with gelatin, albumen, isinglass, bentonite, potassium ferrocyanide or salts of phytic acid; the two last treatments are intended to reduce the level of soluble iron complexes that would otherwise cause 'blackening' of the wine (qv), but with fruit wines they are frequently inadequate. After fining, the wine would normally be filtered through a layer of kieselguhr or through cellulose pulp. Traditional filtration processes through asbestos are nowadays considered unsafe procedures. Alternative clarifying procedures include chilling the wine prior to or after fining and the use of microfiltration systems.

The wine, after clarification, will normally be either flash pasteurized and hot filled into the bottle or will be treated to give a residual SO_2 content. In certain countries, bright wine may be treated with up to 200 mg/l of dimethyl pyrocarbonate (DMPC). This antimicrobial breaks down on contact with water to give reactive products that destroy any low level of microorganisms and leaves only minimal residues of methanol and CO_2; however, as with diethyl pyrocarbonate, which is now banned, there are some concerns about the extended safety-in-use of DMPC because of the production of methyl carbamate in DMPC-treated wines (Sen, Seaman and Weber, 1992; Sen et al., 1993).

Defects of fruit wines. Reports of 'food poisoning' following consumption of 'home-made' elderberry wines were believed to have been caused by high histamine levels. Pogorzelski (1992) investigated the potential for production of histamine in Polish elderberry juices and wines. He found higher levels of histamine in depectinized and pulp-fermented juices than in juices from hot macerated fruits. However, these levels were low by comparison with those from the elderberry fruit wines themselves, in which concentrations up to 12.36 mg/l were found, dependent upon the method of fruit treatment and the yeast strain used for fermentation. The elderberry contains high levels of the amino acid histidine, which serves as substrate for the enzyme histidine decarboxylase. This enzyme is present in the fruit itself and also in certain preparations of pectolytic enzymes and can be produced by many fermentation yeasts (Pogorzelski, 1992).

Flavour defects include taints caused by the presence of hydrogen sulphide, acetic acid, mouse, diacetyl, bitter compounds, yeasty and mouldy taints. Microbial spoilage *per se* results generally from growth of film yeasts such as species of *Hansenula*, *Brettanomyces*, *Pichia* and *Torulopsis* within the bottled wine, growth of acetifying bacteria and (rarely) growth of anaerobic bacteria. The last cause 'bittering' but this is usually restricted to low acid wines.

Turbidity can result from the growth of fermentative yeasts, which will also increase the CO_2 level and may cause bottles to burst. Alternately, turbidity may be caused by chemical precipitation of residual tannins, pectins or proteins (especially in the presence of traces of copper). Discoloration can arise from blackening caused by the precipitation of iron–tannin complexes, browning is caused by oxidation of tannins and anthocyanin pigments, and a bluish-white cloud is caused by precipitation of ferric phosphate. The development of browning as a result of polymerization of anthocyanin pigments in blackberry and raspberry juice wines can be avoided by fermentation of flash-pasteurized depectinized juice in the presence of 50 mg/l sulphur dioxide (Rommel et al., 1992; Withy et al., 1993).

5.4.4 Fruit pulp fermentations

Fermentation of pulp permits a higher level of extraction of anthocyanins and other pigments from fruits such as bilberries, cherries, strawberries and elderberries and facilitates the extraction of juice from most fruits, particularly from plums. A consequence is that the final clarification of the wine is simplified.

Because of the high level of yeasts on the fruit, it is necessary to ensure the addition of a vigorous pure yeast inoculum to compete effectively against the wild yeast population. Since the pulp is a richer source of nutrients than the juice, a vigorous fermentation occurs and nutrient supplementation is not necessary. Natural fermentation can be effective but often produces unpleasant or unsatisfactory flavours. Ideally, the fermenting pulp is held beneath the surface of the fermenting liquid in special tanks in order to exclude the pulp from contact with air. If such tanks are not used, the development of CO_2 in the fermenting pulp will raise the pulp to the surface where rapid oxidation of pigments will occur together with growth of oxidative yeasts. At the completion of fermentation, the wine is racked off and the fermented pulp is then pressed to give an alcohol-rich must that is immediately sweetened and re-fermented, thus increasing the yield.

5.4.5 Alcohol-fortified wines

Fortification of fruit wines with rectified alcohol, of agricultural origin, is practiced in a number of countries to increase the overall alcohol level generally up to 15% ABV but sometimes as high as 20% ABV. Usually this process is undertaken towards the end of, or immediately following, the fermentation period and before maturation. In most instances, the fruit wine is fermented without chaptalization (i.e. addition of fermentation sugar) to an alcohol level of 5–6% ABV and the alcohol is added before maturation in order to permit the 'harmonization' of the added alcohol and the wine. The final product is sweetened by addition of sugar following clarification.

5.4.6 Sparkling (carbonated) fruit wines

As with the production of sparkling ciders, fruit wines may be carbonated either by injection of CO_2 (to a pressure of c. 6 kg/cm^2) or by a *cuvée close* secondary fermentation in a pressurized closed tank. The wine is sweetened by the addition of a small quantity of sugar syrup and inoculated with a suitable yeast culture. The process retains all the developed CO_2 in the wine, which will have a pressure of some 4–5 kg/cm^2. The wine will be filtered and then bottled using an isobaric bottling machine. Critical to the process is the selection of the yeast to be used, which should not generate any specific flavour characteristics. Strains of yeast used for secondary fermentation of champagne are very suitable for this purpose.

5.5 Fruit spirits and liqueurs

The distillation of wines, fruit wines, cider and perry to produce fruit spirits is practised widely. Whilst the best known are brandies such as cognac and armagnac derived from grape wines, traditional spirits such as cider brandy from cider, poire from perry, slivovitz from plum wine and kirsch from black cherries have been produced for many years.

Also widely practised is the alcoholic infusion or decoction of whole or macerated fruits, sometimes with the addition of herbs, spices or other plant material to produce apéritifs, liqueurs and other strong alcoholic beverages with fruit flavour and aroma.

5.5.1 Fruit spirits (sic fruit brandies)

Wine fermented from fruits, or in some instances fermented fruit pulp, is double distilled to produce a spiritous beverage containing up to 70% ABV although in many cases only a single distillation is used, giving a product with an alcohol level of 25–55% ABV. The methods of production vary from area to area but share many common features. Traditionally, distillation takes place soon after completion of fermentation (i.e. without any significant maturation period) in a copper pot still heated over an open flame or in a 'Charentais'-type double distilling still. Generally, the wine will not have been filtered and will contain a small amount of yeast and other sediments. In a two-stage process, the first distillate will contain about 28–30% ABV. During the second distillation, the 'heads' and 'tails' containing high concentrations of aldehydes and fusel oils, respectively, are separated. The middle fraction will have an alcohol content of c. 70% ABV.

The distilled product may be stored in used oak or larch barrels for several years in order to mature and develop a desirable bouquet. However, some fruit brandies (e.g. kirsch) are not stored in wood in order to maintain the clear colourless appearance. For retail sale, the distilled products are generally diluted to an alcohol content of 35–40% ABV or are blended with fruit juices to yield fruit liqueurs (qv).

The quality of the product will be influenced by the quality and variety of the fruit used for the initial fermentation, the pH of the juice or the pulp, the choice of fermentation yeast, the extent to which depectinization was undertaken and the level of use of sulphur dioxide in the primary fermentation.

A major problem with some fruit spirits, such as plum brandy, relates to the level of methanol liberated into the fruit pulp by depectinizing enzymes and subsequently distilled into the final product. Paunovic (1991) reported methanol levels up to 15 mg/l in pectinase-treated plum pulp, with residual levels of methanol in the distilled alcohol of up to 12.4 mg/l alcohol. He

Table 5.9 Some fruit spirits and brandies

Type	Synonym	Country of origin	Fruit	Special characteristics
Alise	Sorbier	France	Rowan berries	
Aprikosengeist	Apricot brandy	Germany	Apricots	
Calvados	Cider brandy	France	Apple	Double distillation from fermented pulp; aged in wood
Eau-de-vie-de-Cidre	Cider brandy	France	Apple	Distilled from cider (cf. Calvados); not usually aged
Kirsch		Germany	Black cherry	Distilled from fermented black cherry pulp
Mirabelle	Plum brandy	France	Yellow Plum	Distilled from fermented plum pulp
Poire Williams		Switzerland, Germany, France, Italy	Williams Pear	Distilled from fermented pear pulp; not aged in wood
Slivovitz	Plum brandy	Former Czechoslovakia, Hungary, Bulgaria	Plum	Distilled from fermented plum pulp; aged in wood

also showed that the level of methanol in the pulp could be reduced significantly by high-temperature treatment of the pulp and careful selection of the fermentation yeast strain: products of 'spontaneous' fermentation tended to have high methanol levels.

Examples of some fruit spirits and brandies are presented in Table 5.9.

5.5.2 Apéritifs and liqueurs

Many well-known brands and styles of fruit-flavoured liqueur are marketed world wide; such products are often produced by 'secret' recipes. Additionally, a number of traditional 'country' recipes are used widely. Some examples are presented in Table 5.10. Normally the process involves the extraction of macerated fresh, dried or fermented fruit pulp with either a neutral distilled agricultural alcohol or with a spirit distillate obtained from the specific fruit wine. However, some products are compounded by blending fruit spirit or neutral spirit with fruit juice, often with added sweetening. Most products have alcohol contents in the range 20–28% ABV.

Gorinstein et al. (1993) investigated the effects of processing variables on the characteristics of a liqueur prepared from persimmon. They observed that the organoleptic qualities (aroma and flavour) of the liqueur could be influenced significantly by the condition of the fruit extracted; the relative ratios of fruit and alcohol; the concentration of alcohol; the extent to which the fruit had been fermented before extraction; and the quality of

130 FRUIT PROCESSING

Table 5.10 Some fruit liqueurs and apéritifs

Type	Country of origin	Fruit	Alcohol type	Alcohol content (% ABV)	Special characteristics
Amaretto	Italy	Apricot (*Armeniaca vulgaris*) and kernels	Eau-de-vie	24	Distillation of fermented fruit pulp then infusion of macerated fruits, kernels and sugar
Cerasella	Italy	Cherry (*Prunus cerasus*)	Neutral spirit or brandy	28	Infusion of macerated fruits; clear, white
Karpi	Finland	Cranberry (*Vaccinium macrocarpon*)	Neutral spirit	30	Infusion of macerated fruits
Lakka, Suomuurain	Finland	Cloudberry (*Rubus chamaemonas*)	Neutral spirit	30	Infusion of macerated fruits and sugar
Rabinowka	Eastern Europe	Rowanberry (*Sorbus aucoparia*)	Neutral spirit	25	Infusion of macerated fruits and sugar
Sloe gin	England	Sloe berry (*Prunus spinosa*)	Gin	22–25	Infusion of whole or macerated fruits and sugar
Cider Royal	England	Cider apple	Cider brandy + cider	20	Blending of cider brandy and cider or apple juice
Reishu	Japan	Melon (*Cucumis melo*)	Neutral spirit	15	Infusion of macerated fruits
Wisniak, Wisnioka	Poland	Cherry (*Prunus cerasus*)	Vodka	25–30	Distillation percolation of alcohol; sweetened
Jerzynowka	Poland	Blackberry (*Rubus fructicosus*)	Neutral spirit or eau-de-vie	25–30	Infusion of macerated fruits
Cassis	France	Black currant (*Ribes nigrum*)	Eau-de-vie	15–20	Distillation of fermented pulp before infusion of macerated fruits; sweetened
Zolataya Osen	Mount Caucasus	Damson (*Prunus spinosa*), apple (*Pirus malus*) and quince (*Pirus cydonia* Lin)	Neutral spirit	20–25	Distillation percolation of alcohol; sweetened

the distilled spirit. The method of choice was to extract macerated, freeze-dried fruits with a middle distillation fraction containing 52–86% ethanol, which was obtained from a 21 day 20–24°C *S. cerevisiae* var. *ellipsoides*-fermented pectinase-treated fresh fruit pulp. The final product contained 27% ABV, 30% total sugar and 30% persimmon extract.

An example of a traditional UK product is sloe gin, which is produced by steeping chopped, destoned sloe (wild plum; *Prunus spinosa*) fruits, gathered from the hedgerows, in gin for a period of 3–4 months. A typical domestic recipe would require 5 kg fruits, 1.2 kg sugar, 25 ml almond essence and 12 l of gin. After steeping, the concoction is filtered and bottled, usually after standardization of the alcohol content to a level of 22–25% ABV. The liquor has a medium to dark red–purple coloration and is a favorite winter drink in the UK, widely associated with field sports. If prefered, the sloes may just be pricked with a large needle; this then permits the extracted fruit to be stored under a layer of fresh gin and consumed as a liqueur gin. Variations on the recipe include the substitution of vodka for gin and use of damsons or other bitter stone fruits.

5.6 Miscellany

The production of fruit-based alcoholic drinks provides a broad range of product types with variations in alcohol content from <1.2% to 55% ABV or more. It is not surprising that local specialities prevail in different countries and regions; it is also not surprising that many diverse products are marketed in the international drinks market. The composition of fruit makes it eminently suitable for the preparation of alcoholic products.

Worthy of consideration, however, are the opportunities which the new era of biotechnological processsses present in this area. Opportunities, for instance, to develop special strains of fermentation yeasts by genetic manipulation: yeasts with the capacity to carry out fermentations at extremes of temperatures, to metabolize fruit polysaccharides (starch) and oligosaccharides, or to produce specific metabolites in order to generate selected flavour profiles. Other opportunities include the use of immobilized yeasts and malo-lactic bacteria to undertake continuous fermentations of fruit juice substrates.

In tropical countries, attention is already drawn to the development of new alcoholic products from indigenous fruits (see section 5.5.1, above). In addition to their use as beverages, such developments include the generation of locally produced fuel alcohol (gasohol) based on fermentation of fruit residues, fruit processing effluents and other fruit wastes. Work in such areas has been reported from Mexico (López-Baca and Gómez, 1992), Japan (Tanemura *et al.*, 1994) and other countries. Fermentation of wastes can also lead to the establishment of local industries for the production of enzymes, food colours, flavours, antimicrobials, etc. (see, for instance, Cook, 1994). All such developments will benefit local and national economies.

An interesting approach which is worthy of further development is the production of fermented *non*-alcoholic beverages enriched in vitamins and other nutrients. Lebrun (1991) has patented a process to produce low-alcohol (<1% ABV) petillant fruit-flavoured drinks by fermentation of a mixture of fruits, plant extracts and sugar.

Of greater interest is the patent of Chavant and Rollan (1992). These workers have developed a process to produce non-alcoholic fruit-based acid drinks enriched with microbial polysaccharides, folic and orotic acids, and a diverse range of B-group vitamins. The novelty arises from the sequential fermentation by selected strains of microorganisms, including the yeasts *Pichia membranaefaciens* and *Candida pseudotropicalis*, and bacteria including *Lactococcus lactis*, *Lactobacillus fermentum*, *Lactobacillus raffinolactis*, *Bifidobacterium* spp., *Leuconostoc* spp. and *Acetobacter* spp. Essentially, they have taken the technology used for many years in the production of the fermented dairy product 'kefir' and applied it to develop a new concept in food beverages.

Such studies suggest potential for the development of new food and non-food products from the fermentation of fruit and their juices.

Acknowledgements

I am indebted to my colleague Ms Lorraine Boddington for assistance with database searches and literature acquisition.

References

Adesina, A.A., Oguntimein, G.B. and Obisanya, M.O. (1993). Kinetic analysis of the fermentation of mango juice. *Lebensmitel-Wissenschaft und Technologie*, **26**, 79–82.

Anon (1994a). European Parliament and Council Directive 94/35/EC of 30 June 1994 on Sweeteners for use in foodstuffs. *Official Journal of the European Communities*, **37**, L237, 3–12.

Anon (1994b). European Parliament and Council Directive 94/36/EC of 30 June 1994 on Colours for use in foodstuffs. *Official Journal of the European Communities*, **37**, L237, 13–29.

Anon (1995). European Parliament and Council Directive 95/2/EC of 25 March 1995 on Additives other than Colours and Sweeteners for use in foodstuffs. *Official Journal of the European Communities*, **38**, L61, 1–40.

Beech, F.W. (1993). Yeasts in cider-making. In *The Yeasts*, Vol. 5, *Yeast Technology* (ed. A.H. Rose and J.S. Harrison), pp. 169–214. Academic Press, London.

Beech, F.W. and Carr, J.G. (1977). Cider and perry. In *Economic Biology*, Vol. 1, *Alcoholic Beverages* (ed. A.H. Rose), pp. 139–313. Academic Press, London.

Beech, F.W. and Davenport, R.R. (1983). New prospects and problems in the beverage industry. In *Food Microbiology: Advances and Prospects* (ed. T.A. Roberts and F.A. Skinner), pp. 241–256. Academic Press, London.

Broussous, P. and Ferrari, G. (1990). Procédé de fabrication d'une boisson pétillante faiblement alcoolisée à base de kiwi. *Brevet Europeen No.* 0 395 822.

Chavant, L. and Rollan, S. (1992). Procédé de fabrication d'un nouveau levain glucidique, nouveau levain obtenu, fabrication de boisson fruitée à partir de ce levain et produit obtenu. *Brevet d'Invention Francaise No.* 2 672 614.

Cook, P.E. (1994). Fermented foods as biotechnological resources. *Food Research International*, **27**, 309–316.

di Baggi, V., Ghommidh, C., Navarro, J.-M. and Crouzet, J. (1986). Fermentation alcoolique de la pulpe de mangue. *Sciences des Aliments*, **6**, 407–416.

Gorinstein, S., Moshe, R., Weisz, M., Hilevitz, J., Tilis, K., Feintuch, D., Bavli, D. and Amram, D. (1993). Effect of processing variables on the characteristics of persimmon liqueur. *Food Chemistry*, **46**, 183–188.

Jarczyk, A. and Wzorek, W. (1977). Fruit and honey wines. In *Economic Microbiology*, Vol 1: *Alcoholic Beverages* (ed. A.H. Rose), pp. 372–421. Academic Press, London.

Jarvis, B. (1993a). Cider (hard cider): the product and its manufacture. In *Encyclopaedia of Food Science, Food Technology and Nutrition* (ed. R. Macrae, R.K. Robinson and M.J. Sadler), pp. 979–983. Academic Press, London.

Jarvis, B. (1993b). Cider (cyder; hard cider): chemistry and microbiology of cidermaking. In *Encyclopaedia of Food Science, Food Technology and Nutrition* (ed. R. Macrae, R.K. Robinson and M.J. Sadler), pp. 984–989. Academic Press, London.

Jarvis, B., Forster, M.J. and Kinsella, W. (1995). Factors affecting the development of cider flavour. In *Microbial Fermentations: Beverages, Foods and Feeds*, (ed. R.G. Board, D. Jones and B. Jarvis). *J. appl. Bact. Symp. Supplement* 1995, **79**, 55–185.

Joshi, V.K., Bhutani, V.P., Lal, B.B. and Sharma, R. (1990). A method for preparation of wild apricot (chulli) wine. *Indian Food Packer*, **Sept–Oct**, 50–55.

Lebrun, F.-A.-L. (1991). Boisson naturellement pétillante et son procédé de fabrication. *Brevet d'Invention Francaise*, No. 2 651 240.

López-Baca, A. and Gómez, J. (1992). Fermentation patterns of whole banana waste liquor with four inocula. *Journal of the Science of Food and Agriculture*, **60**, 85–89.

Maloney, S. (1993). Effects of sulphite in yeasts with special reference to intracellular buffering capacity, PhD Thesis, University of Bath, UK.

Millard, P.S., Gensheimer, K.F., Addiss, D.G., Sosin, D.M., Beckett, G.A., Houch-Jankoski, A. and Hudson, A. (1994). An outbreak of cryptosporidiosis from fresh-pressed apple cider. *Journal of the American Medical Association*, **272**, 1592–1596.

Mosso, K., Aboua, F., Angbo, S., Konan, K.E., Nyamien, M.N. and Koissy-Kpein, L.M. (1990). Evolution des sucres totaux, de l'extrait sec réfractométrique et de l'alcool, au cours de la fermentation vineuse à température alternées de la banane poyo, de la banane plantain et de la papaye. *Actualités des Industries Alimentaires et Agro-alimentaires*, **Sept**, 767–772.

Nagy, S., Chen, C.S. and Shaw, P.E. (ed.) (1993). *Fruit Juice Processing Technology*, Agscience, Auburndale, FL.

Paillard, N.M.M. (1990). The flavour of apples, pears and quinces. In *Food Flavours*, Part C. *The Flavour of Fruits* (ed. I.D. Morton and A.J. Macleod), pp. 1–41. Elsevier Science, Amsterdam.

Paunovic, R. (1991). Uticaj izazivaca i uslova izvodenja alkoholne fermentacije covnog kljuka na sastav vocnih rakija. *Archiv za Poljoprivre Nauke*, **52**, 186, 171–183.

Pogorzelski, E. (1992). Studies on the formation of histamine in must and wines from elderberry fruit. *Journal of the Science of Food and Agriculture*, **60**, 239–244.

Pollard, A. and Beech, F.W. (1963). The principles and practice of perry making. In *Perry Pears* (ed. L.C. Luckwill and A. Pollard), pp. 195–203. University of Bristol.

Poullet, H. (1994). Procédé de fermentation de moût de Carambole. *Brevet d'Invention Francaise*, No. 2 695 136.

Rommel, A., Wrolstad, R.E. and Heatherbell, D.A. (1992). Blackberry juice and wine: processing and storage effects on anthocyanin composition, color and appearance. *Journal of Food Science*, **57**(2) 385–391, 410.

Rose, R. (1991) Boisson alcoolique pétillante à base d'orange. *Brevet d'Invention Francaise*, No. 2 657 878.

Sen, N.P., Seaman, S.W. and Weber, D. (1992). A method for the determination of methyl carbamate and ethyl carbamate in wines. *Food Additives and Contaminants*, **9**, 149–160.

Sen, N.P., Seaman, S.W., Boyle, M. and Weber, D. (1993). Methyl carbamate and ethyl carbamate in alcoholic beverages and other fermented foods. *Food Chemistry*, **48**, 359–366.

Swaffield, C.S., Scott, J.A. and Jarvis, B. (1994). Influence of selected microbial isolates on development of mature cider flavours. *Proceedings of the Society for Applied Bacteriology Summer Conference 1994. Journal of Applied Bacteriology*, **77** (Suppl.), xi.

Tanemura, K., Kida, K., Ikbal, Matsumoto, J. and Sonada, Y. (1994). Anaerobic treatment of wastewater with high salt content from a pickled plum manufacturing process. *Journal of Fermentation and Bioengineering*, **77**, 188–193.

Teotia, M.S., Manan, J.K., Berry, S.K. and Sehgal, R.C. (1991). Beverage development from fermented (*S. cerevisiae*) muskmelon (*C. melo*) juice. *Indian Food Packer*, **July–Aug**, 49–55.

Tjomb, P. (1991) Boissons au kiwi: c'est l'effervescence. *Revue de l'Industrie Agro-Alimentaire No. 468*, **Oct**, 42, 44–45.

Williams, R.R., Faulkner, G. and Burroughs, L.F. (1963). Descriptions of the principal varieties of perry pears. In *Perry Pears* (ed. L.C. Luckwill and A. Pollard), pp. 61–172. University of Bristol, UK.

Withy, L.M., Heatherbell, D.A. and Fisher, B.M. (1993). Red raspberry wine – effect of processing and storage on colour and stability. *Fruit Processing*, **3**, (8), 303–307.

Yang, H.Y. and Wiegand, E.H. (1949). Production of fruit wines in the Pacific Northwest. *Fruit Products Journal*, **29**, 8–12, 27, 29.

6 Production of thermally processed and frozen fruit
G. BURROWS

6.1 Introduction

Most fruits have a definite harvesting season, usually of very short duration, meaning that although there is an abundant supply during this period it is not available at other times of the year. The fruit itself has a limited keeping time. Modern forms of transport make available fruits that are not in season locally (e.g. strawberries to Britain in February) but this is expensive.

Fruit quickly deteriorates, the two main causes being microorganisms and biochemical activity. It is living and respiring when it is harvested, with the biochemical processes catalysed by a number of enzymes. In order to preserve the fruit, the enzymes must be inactivated or the fruit must be preserved in some other way. However, even when biochemical activity has been stopped, microbial infection can still occur and cause further deterioration.

So that fruit can be easily available all year round, different methods of preservation have to be applied. These methods often alter the characteristics of the fruit to a greater or lesser degree. Different methods may extend the shelf life by a few weeks while others can give a shelf life of two years or more. The methods commonly used for long-term preservation are canning, bottling and freezing, while aseptic preservation is used only in a limited way for particulate fruits. Each of these methods has both advantages and disadvantges and these will be discussed in this chapter.

The choice of method of preservation may well be determined by the raw material. Varieties that are suitable for canning with the addition of artificial colour may not be suitable for freezing where no added colour can be used.

6.2 Raw materials

The raw material must be sound, ripe, free from blemishes and disease. If the raw material is unsuitable for heat processing because it is of the wrong variety or it is underripe, the final product may have a poor colour, texture and flavour. Victoria plums that are not fully ripe and still have a green colour will not give a satisfactory product however much colouring matter is used because the natural chlorophyll turns brown during the heat process required for canning. Some varieties of pears acquire a pink coloration after heating, particularly if they are not thoroughly cooled. The chemistry of this reaction is not fully understood, but it is believed to be caused by the natural pigments in the fruit combining with any metal ions to form complexes.

Some fruits, such as apples and pears, tend to go brown when peeled and exposed to the air, thus causing a poor colour prior to being processed. The browning effect can be prevented by placing the fruit in dilute brine or a solution of ascorbic acid.

Plums and gages sometimes turn brown when the fruit is exposed to the headspace between the surface of the liquor and the lid of the can, but if this part of the process is properly controlled by eliminating as much air as possible from the headspace by use of steam flow closure, discoloration can be prevented.

Certain soft fruits, such as strawberries and some varieties of raspberries, lose their colour during heat processing and may produce brown hues. Artificial colours can be used in these instances but new varieties have been developed that will retain their colour. This is particularly important where fruit is canned in fruit juice with no other additives.

6.3 Canning of fruit

6.3.1 Cannery hygiene

High standards of hygiene are required to avoid losses through the product being spoilt or contaminated by debris on the line or by pests attracted by the debris. The schedules for cleaning and housekeeping must be documented and operated by properly trained staff. Equipment should be selected and sited with ease of cleaning in mind.

Cleaning should be monitored both visually and microbiologically. The former can be carried out as part of a preproduction check by the line supervisor, while there are rapid luminescence methods available to assess quickly the microbial contamination.

Conveyors, pipelines and any surfaces that come into contact with food should be constructed from smooth, durable materials and should be as free as possible from crevices and inaccessible corners.

Cooling water and all post-process handling equipment require special attention to avoid the possibility of post-process spoilage.

All storage and production areas must be properly screened to prevent ingress by pests such as rodents, birds and insects, which can cause damage to raw materials. It is advisable to engage the services of a specialist pest control company to ensure that all protective measures have been employed.

6.3.2 Factory reception

Fruit is delivered to the factory by lorry or, if in small amounts, by trailer, usually in crates, boxes or small punnets, depending on its susceptibility to damage.

Each load is weighed on arrival to determine the payment to be made to the supplier. Samples are taken from each delivery to ensure the quality is suitable for processing. Some factories operate a sliding scale of reduction of payment if the fruit quality is not quite right. The sampling may reveal that the fruit is too ripe to be stored and that it must be canned immediately before further deterioration takes place. If foreign material, such as glass, is found, the load may be rejected.

Once accepted, the fruit must be canned with the minimum of delay, particularly if it has been picked during the heat of the day. Deterioration and mould growth will occur quickly in warm fruit. Ideally, fruit should be canned on the day of picking and many factories during the fruit season work a shift pattern that goes into the night to ensure that this happens.

If fruit does have to be stored for any time, it should be placed in a cool dry environment. Care must be taken to prevent the fruit being attacked by birds, insects or other pests. Chilled stores operating at 5–8°C may be used.

6.3.3 Peeling

Methods used for peeling fruit vary from hand peeling to a variety of mechanical methods; in all cases there will be a need for visual inspection and some hand trimming to remove any remaining skin or blemishes. It is possible to peel these away mechanically but it is very wasteful.

Abrasion peeling, on its own, tends to be wasteful as the whole fruit has to be 'rubbed' down to remove the skin and blemishes. It is more simple to control peeling losses by using lye (caustic soda). The fruit is passed through a hot solution of caustic soda (sodium hydroxide, NaOH) and the degree of peeling can be adjusted by varying the strength and the temperature of the solution and the residence time. In practice the temperature is usually maintained at, or close to, boiling point. The contact time may vary from 1 to 2 minutes in 2% to 10% solution. The loosened skin is removed by jets of water, which also remove all traces of caustic soda. Great care must be taken to avoid accidents with boiling caustic soda solutions.

Some fruits, particularly apples and pears, may be knife peeled by machines on which the fruit is impaled and rotated against a knife that follows the contours of the fruit. The core may be removed at the same time. These machines tend to be wasteful, particularly if the fruit has not been carefully size-graded.

Flame peeling has been used for apples, with the loosened skin being removed by mild abrasion under jets of water. Peeling losses are low, but great care is needed to ensure that all charred skin is removed as this is very unsightly in the finished pack.

6.3.4 Blanching

Some fruits may require blanching prior to filling, particularly if they are to be solid packed, since the softening and shrinkage enables them to be more readily filled into the cans. The hot filling also reduces the processing time where heat penetration may be slow, but, conversely, cooling of the contents of large cans may be very slow, with consequent loss of quality.

The advantages of blanching may be slightly offset by the loss of nutritional values that occurs during the operation. Therefore, blanching times should be kept as short as possible. The nutritional-value loss can be minimised by blanching in steam rather than in water, as the leaching losses are lower.

6.3.5 Choice of cans

It is preferable to use plain bodied cans for some fruits, such as apples, as these help to keep the colour and flavour bright and fresh. This is the result of a chemical reaction between the fruit and the tinplate. The presence of tin gives the fruit a brighter colour. However, other fruits, such as rhubarb and plums, must be filled into lacquered cans to prevent the acid in the fruit reacting with the tinplate.

6.3.6 Filling

Irrespective of whether the cans are filled by machine or by hand, regular checks must be made on the temperature of the product at the time of closing the can, as this may affect the subsequent exhausting and pasteurising processes. The filled weights must also be controlled. This may need to include the weights of the constituent parts of the pack, e.g. fruit(s) and syrup, in order to ensure compliance with recipe requirements or legal standards. Correct filling is not only desirable for economic reasons but also technically important.

All weight control data and temperatures should be recorded on charts kept near to the filling area so that trends can be observed quickly and any necessary corrective action taken.

6.3.7 Syrup

Apart from some solid pack products, fruits are usually canned in a sugar syrup, although there is an increasing use of fruit juice in these more health-conscious days. The syrup is usually prepared from granulated sugar obtained from beet or cane but it is possible to use other sugars, such as dextrose, corn syrup, glucose syrup or invert sugar. Sugar strengths are normally expressed in degrees Brix (°B) which is a measure of the percentage of sugar by weight in an aqueous solution at 20°C.

The UK Canners Code of Practice defines light syrup, syrup and heavy syrup. These vary according to the type of fruit, a distinction being made between the following three classes:

Class A: apples (other than purée or solid pack), bilberries, blackberries, black currants, cherries, damsons, gooseberries, greengages, loganberries, plums, raspberries, redcurrants, rhubarb and strawberries
Class B: apricots, peaches, pears, pineapple, fruit salad and fruit cocktail
Class C: prunes.

Table 6.1 details the different syrup strengths for the different classes of fruit.

Sugar density can be checked by using a direct Brix reading hydrometer. If the temperature is other than 20°C then a correction factor has to be applied. It is often more convenient to use a refractometer to measure the strength of the syrup. These are available as bench models or as hand-held instruments, which are particularly useful when an instant reading is required.

6.3.8 Cut out

When examining cans of fruit to determine their quality, measurements are made of the drained weights of the fruit and the density of the covering liquor. This examination is known as the 'cut out' and is best carried out after canning and when the contents have reached equilibrium (48 hours or more).

Syrup may be added at 45°B, but when the can is examined it will

Table 6.1 Syrup strengths for different classes of fruit

	Class A (% w/w)	Class B (% w/w)	Class C (% w/w)
Light syrup	15	15	10
Syrup	30	22	15
Heavy syrup	40	30	20

Table 6.2 Drained weights as percentage of filled weights for fruits canned in 40–45°B syrup

Fruit	Drained weight as percentage of filled weight
Blackberries	65–90
Black currants	75–95
Cherries	85–100
Damsons	85–95
Gooseberries	85–100
Loganberries	70–90
Raspberries	65–95
Strawberries	55–80

probably cut out at 25 to 30°B. This is because it has been diluted by the water naturally present in the fruit. The amount of dilution will depend on the type of fruit, its variety and state of ripeness, and on the ratio of the fruit to syrup in the can.

If the weight of the fruit packed in the can is known, it is possible to calculate fairly accurately the strength of the syrup used. The weight of drained solids in any particular can is not constantly proportional to the filled weight, as it can be influenced by a number of factors, such as the exhaust time and temperature, the condition of the fruit and the strength of the syrup.

Canned fruits packed in 40 to 45°B syrup generally give the approximate drained weights expressed as a percentage of the filled weight as shown in Table 6.2. The variations are caused by the differences in texture. If 30°B syrup is used, the drained weights will be approximately 2% less than those shown.

6.3.9 Closing

Cans are closed by placing the lid on a filled can and sealing it to the body by the formation of a double seam (Figure 6.1). As the term implies, the seam is formed by two operations in the seamer. During the first operation the can end seaming panel is rolled together and interlocked with the can body flange. The second operation completes the seam by pressing it to the required tightness. Details of double-seam technology and the methods of measuring the seam dimensions can be found in the can manufacturers' seam manuals.

For a number of reasons, it is important to control the temperature of the can contents at the time of closure. The air enclosed in the can will affect the final vacuum, which, in turn, will influence the shelflife of the product by controlling internal corrosion.

The choice of the closing temperature depends on the type of product

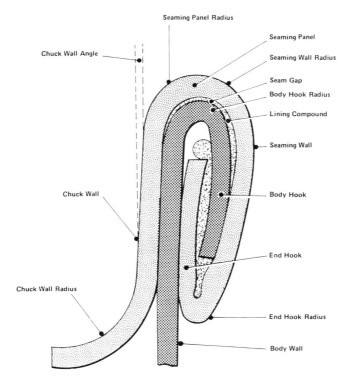

Figure 6.1 Cross section of a can double seam (courtesy of Carnand Metalbox plc).

and the techniques used for preparation, which may preclude hot filling. In these cases, alternative methods of obtaining a vacuum must be employed.

6.3.10 Exhausting

This is the term given to the process of removing air and entrapped gases from the can before closing. It can be achieved by several methods, depending on the type of product.

A product containing a thin liquid will only rarely occlude gases below its surface and, therefore, only requires the removal of the air in the headspace. Viscous or semi-solid products may contain a considerable amount of entrapped air when filled into the can. Fruit and vegetable tissues also may contain CO_2 evolved from the respiratory process.

The syrup used for covering fruits is normally filled as hot as possible (ideally greater than 80°C) so the air in the headspace is partially displaced by the steam from the hot liquid.

Can closing is often preceded by steam-flow closure, in which steam is injected into the headspace from jets on the seamer. This has the effect of sweeping the air from the headspace immediately before the lid is seamed on, thus creating a partial vacuum as the steam condenses.

6.3.11 Can vacuum

In general, the presence of an adequate vacuum in a can is a sign of good canning practice. The exceptions are carbonated drinks, which have a positive internal pressure, and milk products, which have little or no vacuum.

The presence of a vacuum ensures that the can ends are concave, thus allowing a visual means of detecting cans that have an internal pressure resulting from gas formation caused by spoilage.

Fruits, being of a highly corrosive nature because of their acidity, require a high vacuum of 25 mm-g or above, so that any hydrogen formed by corrosion will dissipate into this vacuum. The high vacuum also ensures a low amount of O_2 thus ensuring corrosion is slow.

The vacuum in a can is normally measured with a Bourdon type gauge, which has a sharp tapering hypodermic needle projecting through a thick rubber washer. When pushed with the hand on the can lid the needle penetrates the tinplate while the washer acts as a seal. The reading on the dial is taken before the gauge is removed from the can and the seal broken (Figure 6.2).

6.3.12 Processing

The most important part of the canning process is the destruction of bacteria by heat. Bacteria in their active state are not particularly heat resistant but spores often present considerable heat resistance and, as a consequence, temperatures up to 130°C are required to destroy them.

In order to achieve microbiological stability of canned fruit, it is necessary to submit the sealed can to a heat process that will destroy, or render inactive, all microorganisms capable of causing spoilage. Absolute sterility is seldom achieved and, therefore, sterilisation is not a strictly accurate term. The term processing is generally used to describe the heat treatment.

However, pH plays a part in the heat preservation of fruit. Under acid conditions (values pH less than 3.7) bacteria will not multiply and consequently only a pasteurisation process is necessary. This process is normally carried out by immersing the sealed can in boiling water or steam at atmospheric pressure for relatively short times.

It is not possible to give actual processes for the different products and can sizes but an indication can be obtained from Technical Bulletin No 4 produced by the Campden and Chorleywood Food Research Association.

PRODUCTION OF THERMALLY PROCESSED AND FROZEN FRUIT 143

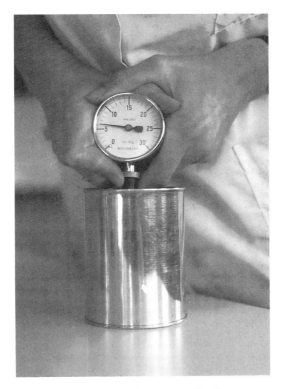

Figure 6.2 Cross section of a can double seam.

For products with a pH value between 3.7 and 4.5, there are a few bacteria that are able to multiply in this range and these products require a longer process or the acidification of the product.

Some fruits, such as prunes, are processed for much longer than is required to inactivate the bacteria, the extra treatment being required to achieve the correct texture of the fruit.

Fruit is normally processed in boiling water or in steam at atmospheric pressure in batch or continuous retorts or in agitating cookers. The basic processing vessel is the static retort, which may be vertical or horizontal. It is equipped to receive steam and water and has suitable drain valves and vents. The cans are loaded into cages, which are placed into the retort and completely covered with the heating medium. If steam is used, care must be taken to vent off all of the air to avoid the possibility of cold spots in the retort, which may cause some cans to be incorrectly pasteurised.

Some batch retorts have a means of rotating the cage in the heating medium, which has the effect of reducing the processing time because the agitation of the contents causes better heat transference.

Continuous-pressure cookers may also be used. These are equipped with specially designed ports or valves that allow the containers into and out of the processing chamber whilst allowing a constant steam pressure to be maintained. There are considerable advantages in using such equipment as there is less variability in the treatment of the cans and much less manual handling. Some continuous cookers cause the cans to rotate against the outer retort casing, thus causing agitation with its attendant advantages (Figure 6.3).

Hydrostatic cookers are sometimes used for fruit, but they require such large quantities of cans to fill the infeed and outfeed legs as well as the heating chamber that they are not always the most cost effective way of processing fruit.

After processing it is important that the cans are correctly and swiftly cooled and dried and that the filled cans are stored under the correct conditions. They should be kept in a cool, dry and clean area, preferably away from direct sunlight. Sudden changes in ambient temperature should be avoided as the can contents may be below the dew point of the atmosphere. This could result in condensation, which will cause rusting of the exterior of the can.

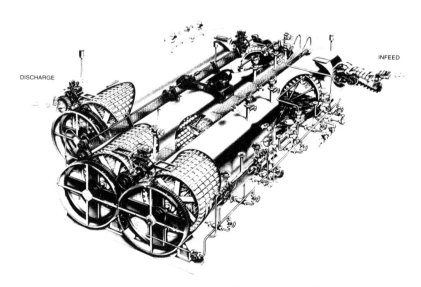

Figure 6.3 A continuous rotary cooker and cooler (Sterilmatic) suitable for processing cans of fruit (courtesy of the FMC Corporation Food Machinery).

6.3.13 Finished pack pH values

The pH of most canned fruits lies between 2.7 and 4.3. The list shown in Table 6.3 is not exhaustive and is laid out in ascending order of average pH value.

6.4 Varieties of fruit

6.4.1 Apples

Apples for canning must be of good size and shape, evenly graded and of good quality. The best English apple for processing is the Bramley Seedling, which is probably the most well-known cooking apple. Newton Wonder has also been canned with some success. Golden Delicious is grown more on the European mainland and will give a product similar to the Bramley Seedling, particularly if the acidity of the covering liquor is chemically adjusted with citric acid.

Cox's Orange Pippin has been used for sweet packs, such as sliced apples in syrup. All the apples used must be of a good evenly graded size and of a good shape, so that the resulting peeled segments are approximately the same size. It is not possible to use windfalls or other damaged fruit, as the removal of the bruised or discoloured sections will impair the quality of the finished pack.

Ideally when the apples arrive at the factory, the quality should be such that washing and grading are unnecessary and the fruit can move to a machine that will core and peel in one operation. After peeling, the apples are dropped directly into a tank of dilute brine (1½%) to prevent them coming into contact with the air, which will cause browning by oxidation or enzyme activity.

The peeled apples are moved quickly through a machine to remove the seed cells and then they are hand trimmed to remove any remaining blemishes. Following this they are cut into sections, usually sixes or eights

Table 6.3 The pH values of finished packs of some fruits

Fruit	pH	Fruit	pH
Rhubarb	3.2–3.6	Damsons	2.9–3.4
Purple plums	2.7–3.3	Blackberries	2.8–3.5
Loganberries	2.7–3.1	Greengages	3.0–3.5
Golden plums	2.9–3.2	Acid cherries	3.1–3.4
Victoria plums	2.8–3.3	Sub-acid cherries	3.2–3.4
Gooseberries	2.7–3.3	Strawberries	3.2–3.8
Apples (solid)	2.8–3.3	Sweet cherries	3.7–4.4

depending on the size of the apple. The sections are kept in dilute brine for a number of hours, sometimes overnight. This improves the texture, colour and keeping qualities of the canned apples and, by removing all the air entrapped in the fruit tissues, renders the can less likely to internal corrosion. If this is not effective then the apples must be gently blanched in a stainless steel blancher.

Immediately prior to use, the apple segments are spread onto a white conveyor belt where they are sprayed with fresh water to remove the brine and inspected to remove any remaining discoloured pieces.

There are a variety of ways of canning apples, some of which are outlined below.

Solid pack. This is by far the most common way of canning apples, often in A10 cans for the catering trade. (See Table 6.4 for a list of the major can sizes used in the UK.) After the inspection procedure, the apples are blanched by passing the sections through a steam tunnel or by immersion in hot water for a sufficient time to render the slices pliable. Filling should be carried out immediately to prevent any appreciable drop in temperature. Ideally the temperature should be above 70°C. The cans are then topped up with boiling water, taking care to maintain the headspace at between 8 mm and 10 mm as any excess air will give rise to internal headspace rusting. After filling, the cans should be thoroughly exhausted to ensure that all air is removed. When the cans have been seamed, they are heat treated in a retort for sufficient time to raise the can centre temperature to above 70°C. This time will be reduced considerably if the cans are agitated during

Table 6.4 Can sizes used for fruits in the UK

Common name	Metric size (mm)	Imperial size	Capacity[a] (ml)
3-piece cans			
Picnic	65 × 78	211 × 301	235
A1	65 × 102	211 × 400	315
U8	73 × 62	300 × 207	230
14Z	73 × 105	300 × 402	400
ET	73 × 110	–	425
UT	73 × 115	300 × 408¾	445
A2	83 × 114	307 × 408	580
A2½	99 × 119	401 × 411	850
A6	153 × 152	603 × 600	2630
A10	153 × 178	603 × 700	3110
2-piece cans			
ET (round)	73 × 110	–	425
Pie can (taper)	153 × 32	603 × 104	450
Bowl	83 × 60	307 × 206	240
Bowl	105 × 71	404 × 213	495

[a]Capacity to nearest 5 ml.

processing. If the contents are hot filled and there is no delay before closing, the cans may just be immediately inverted. Following the heat process the cans are thoroughly cooled.

Apples in syrup. After blanching, the apple segments or, in some cases, the dice are filled into cans and hot syrup is added. The cans are then seamed using steam-flow closure and are pasteurised in boiling water for 10 to 25 minutes, according to can size, following which they are properly water cooled.

Apple sauce (purée). Good-quality apple sauce cannot be prepared from waste produced during canning of apple segments. It is possible, however, to use undersize or oversize fruit that is in good condition. The apples are peeled and sliced or chopped into pieces about 6 mm thick, which are cooked in water or steam until they are soft. The apple is then passed through a pulper with screens of 3 mm or less to remove any seeds or remaining core. Dry sugar is added to the pulp, which is then mixed and re-heated. It is possible to use jacketed pans or a screw heat exchanger, but the best results are obtained using a flash pasteuriser.

If the apple purée has been heated in a continuous screw cooker, it can be filled directly into the cans, the lids seamed on and no further processing is necessary. If the temperature has fallen because the purée has been tipped from the jacketed copper in which it has ben prepared, the cans will require further heat processing after seaming. This is carried out in a retort.

Apple pie. This is produced in a 603 mm diameter tapered pie can. As it requires pressure processing, there are two main problems to overcome. The quality of the filling must be retained while moisture absorption must be prevented to avoid soggy pastry.

A filling will contain apple segments or dice combined with starch and, if so desired, a clove-based spice mix. The starch is necessary to hold the apple in suspension so that there is a homogeneous mixture within the can. Formulation of the pastry is critical and will require much product development to suit a particular process. The pH of the filling must be controlled to keep it below 4.0 by the addition of citric acid.

A similar product, known as apple pudding, may be prepared by placing a slab of pastry into a can, filling with sliced apple with a little syrup and placing a pastry lid on top before exhausting, seaming and processing.

Plain-bodied cans and lacquered ends are used for apple products. Owing to the reaction of the fruit with the tinplate, the plain body gives the apple a bright colour. Fully lacquered cans may impart a dull colour.

It is very important that apples do not come into contact with any iron or copper as this will cause a chemical discoloration of the fruit. All knives, preparation and filling equipment must be made from stainless steel.

6.4.2 Apricots

Most apricots are canned in their country of origin and the most popular variety is Bulidon. They should be delivered to the cannery fresh and must not be overripe.

The fruit is size graded and the stone is removed by splitting the fruit into two halves. The apricot halves are then washed by water sprays and allowed to drain before being filled into the can.

Apricots are rather acid and a heavy syrup (50°B) is recommended. After seaming and pasteurising, the cans are properly cooled before being packed and stored.

6.4.3 Bilberries

These are sometimes referred to as blueberries or blaeberries. They are not processed to a great extent. In some areas this product is in demand by confectioners and in this case the fruits are usually canned in water in A10 cans.

The pack is very corrosive and it is important that good canning practice is meticulously observed and that fully lacquered cans are used.

6.4.4 Blackberries

Cultivated blackberries are used for processing. The most commonly used varieties are Black Diamond and Himalayan Giant. They should be processed when fully ripe as the flavour is at its best at this stage and the seeds are less noticeable. The fruit should be picked into shallow baskets or trays to prevent crushing and it should be processed as quickly as possible after picking as any delay will cause deterioration.

The blackberries are tipped onto an inspection belt where any leaves, calices and other foreign material are removed. They are then washed in a flood washer and sprayed with clean water as they emerge on an inclined belt. Following this they are filled into cans, hot syrup is added and the cans are seamed and pasteurised.

6.4.5 Blackcurrants

Probably the best variety for processing is Baldwin. Only firm ripe fruit should be used. Underripe currents should be avoided as they have tough skins and a poor colour and flavour. Removal of the stalks by hand is both costly and time consuming so, on a commercial basis, strigging is carried out mechanically. It can only be done on fresh currants if they are very firm and dry and it is best carried out on hard frozen fruit. Only small amounts are removed from the cold store at any one time to prevent thawing. After being strigged the fruit is visually inspected to remove any foreign material.

When the frozen fruit is canned, extra exhausting time must be allowed to give the fruit chance to reach normal closing temperature. Heat must be applied fairly gently to prevent the currants from bursting.

6.4.6   Cherries

There are three types of preserved cherries; the sweet white type, the red sour fruit, which is most often pitted, and the characteristically flavoured Maraschino.

The sweet cherry varieties that are most suitable for canning are Napoleon Bigarreau, Kentish Bigarreau and Elton Bigarreau, all being 'white' varieties. These usually require colour to make them red. As these cherries are packed into non-lacquered cans erythrosine is generally used as a colourant because it is relatively stable with regard to dissolved tin. If the cherries are to be used as part of a fruit salad, the amount of colour must be very carefully controlled to prevent subsequent 'bleeding' of the dye, which could be picked up by the other constituents of the pack.

None of the black coloured sweet cherries has been found to be totally satisfactory for canning, although Hertfordshire Black is occasionally used.

The most suitable varieties of sour cherries are Morello (by far the most popular), May Duke, Kentish Red and Flemish Red.

The Maraschino derives its name from a Dalmatian sweet–sour liqueur originally prepared from Amarasco bitter cherries. These are usually preserved as glacé cherries or as bottled cocktail cherries and the best variety is Royal Anne. The cherries are normally received preserved with sulphur dioxide in barrels. If they are to be canned, all of the sulphur dioxide must be leached out or the cans will be severely corroded.

When canned, the cherries should be fairly ripe to give the maximum amount of flesh in relation to stone.

When the fruit arrives at the factory, the stalks are removed by a machine consisting of a series of contra-rotating rollers. The cherries are then inspected and graded prior to being filled into cans.

Some packs are produced from 'pitted' fruits, which have had the stones removed by a pin-like probe that passes through the centre of each cherry, after which the fruit is visually inspected for any remaining stones.

After being filled, the cans are exhausted, processed and cooled.

6.4.7   Gooseberries

Some fruits do not retain the best flavour when canned but gooseberries most definitely improve in flavour. The most important varieties for processing are Careless and Keepsake. They should be picked as they reach full size but before they have become soft or have shown a colour change.

On arrival at the factory, the fruit is passed through a cleaning and grading machine in which the leaves and other light vegetable material are removed by air currents supplied by fans. The stalks and blossom ends are removed in a carborundum peeler or snibber. Care must be taken to ensure that just the stalks and blossom ends are removed and to avoid cutting deeply into the fruit. The fruit is washed with water sprays to remove the debris. The snibbing operation also causes the skin of the gooseberry to be punctured or scarified, which will allow the added syrup to penetrate into the fruit and thus prevent it becoming wrinkled.

After the gooseberries are filled into the can, syrup is added and the can is seamed using steam-flow closure. The cans are then pasteurised for 5 to 30 minutes depending on their size.

This is one of the fruits that benefits from being packed in plain-bodied cans. Use of lacquered cans gives a dull looking pack whereas the plain cans result in a bright attractive pack.

6.4.8 Grapefruit

Grapefruit for canning must be ripe, sound and size graded, ideally between 7.5 cm and 10.2 cm diameter. Any smaller or larger fruit are diverted to produce juice. It is uneconomic to process small fruit owing to the large percentage of peeling losses and other waste. Large fruit give too few segments in a can.

The most difficult aspect of the preparation of the fruit for canning is the separation of the segments from the peel and the white surrounding membrane. Most of the bitter 'principle', naringin, is found in the white membrane, which must be completely removed.

Grapefruit are peeled by one of two methods. In the cold peel method, the peel is stripped off manually to remove both albedo and the outer membrane of the segments at the same time. This method results in a low yield and gives segments that occasionally show signs of overtrimming. The more common method is hot peeling in which the grapefruit are scalded by steam or in water at about 95°C for about 5 minutes to make the peel plump and puffy and easy to remove. After slight water cooling, the peel is removed by hand using a short stainless steel knife. The stem end of the fruit is slit or scored and the grapefruit is popped out of the peel. Some of the albedo or inner white portion of the peel adheres to the fruit and this is stripped away.

There are mechanical peelers that successfully remove the skin from the fruit. The grapefruit are treated for about 12 seconds with 0.5% to 2.5% caustic soda solution at 95°C to dissolve any adhering albedo. The caustic solution is washed off with water sprays.

The major disadvantage of the high-temperature method is that the heat causes the fruit sacs to expand and these may eventually burst.

The peeled grapefruit is sectioned either mechanically or by hand, the seeds removed and the sections filled into cans. Syrup is added and the can end seamed on.

6.4.9 Fruit salad

This is the largest fruit pack canned out of the normal fruit season. Because it is produced as an out-of-season product and also because the fruits used originate from a number of sources, it consists largely of re-canned fruits. Special packs of the individual fruits are canned into A10 or larger cans to be re-used later. The raw material has to be carefully selected to withstand the double amount of heat processing it will receive. This is particularly important for pears and apricots, which must be properly ripe but not so soft that they will break down during the process.

The British Fruit and Vegetable Canners Association (BFVCA) Code of Practice states that canned fruit salad should contain fruits in the following proportion of the filled weight of fruit: peaches 23–46%, apricots 15–30%, pears 19–38%, pineapple 8–16%, cherries or grapes 5–15%.

The syrup drained from the canned fruits may be used in the preparation of the syrup to be used in the fruit salad after it has been filtered and adjusted to achieve the specified strength.

The fruits are usually prepared to give approximately the same count of each in the can. In order to achieve this, it is necessary to include peach and pear quarters, apricot halves, pineapple slices and whole cherries.

The cherries used for this pack must be carefully dyed with erythrosine. Care must be taken not to use too much colouring as this may 'bleed' in the finished pack thus passing on a pink tinge to the other fruits, particularly the white fleshed pears and the pale yellow pineapple. As an added precaution, it is advisable to fill the pears first into the can, followed by the pineapple, apricots and peaches, with the cherries being added last. This will prevent the pears picking up the colour.

6.4.10 Fruit cocktail

This is similar to fruit salad but is prepared with diced fruit. The BFVCA Code of Practice states that fruit cocktail should consist of: diced peaches 30–50%, diced pears 25–45%, pieces or diced pineapples 6–25%, cherries 2–15% and, sometimes, seedless grapes 6–20%.

Both fruit salad and fruit cocktail are best packed into plain-bodied cans with lacquered ends, which will result in a final pack with a bright colour and a true sharp flavour.

6.4.11 Fruit pie fillings

This is a range of products that has become quite popular in recent years. It can be used hot as a filling in a pie or it can be used cold as a dessert topping or accompaniment. The fruit, either single or mixed, is prepared in the normal manner. For convenience, frozen fruit is often used. It is blended with a syrup that contains 3–4% starch and about 40% sugar, together with added colour if appropriate. When fruit is defrosted, it has a high enzyme activity. This can be minimised by accelerating the thawing process, either in a tank or by a continuous process, and pasteurising the fruit shortly after thawing is completed. The use of a thermoscrew heat exchanger is ideal for this purpose.

Owing to the solid nature of the can contents, which results in a slow rate of heat penetration, it is necessary to give pie fillings a longer process than particulate fruit in syrup, juice or water. Many of the pie fillings are hot filled and this will enable a reduction of the processing time.

6.4.12 Loganberries

These originated as a cross between a wild blackberry and a raspberry and are reasonably successful as a processed pack. They have to be handled carefully in a similar manner to raspberries to prevent them bursting, with a resultant loss of juice.

The fruit should be packed ripe but not black. It must not be underipe otherwise the core of the fruit, which is tough, will be difficult to cook. An underripe berry in a can, because of its greater resistance to heat transfer, could retain gas-producing microorganisms in the core that have survived the normal heat treatment. These microorganisms could then cause the can to swell.

Loganberries should be canned immediately on arrival at the factory as they tend to go mouldy very quickly. They should be inspected carefully for maggots and any suspect fruit must be removed.

When loganberries are filled into the can, they must not be pushed down tightly as they easily crush.

6.4.13 Oranges

This pack is canned on the European continent from fresh, ripe sound fruit.

The washed oranges are passed through a steam or hot water bath for about 1 minute to aid peeling, following which they are cooled in cold water. They are then peeled by hand and allowed to dry before being segmented by hand. The individual segments are immersed in cold 2.5% hydrochloric acid for about 2 hours to remove the stringy fibres adhering to them. The acid is subsequently removed by running water and the washed segments are immersed for about 20 minutes in a 1% solution of caustic soda before being washed again.

The separated orange segments are very fragile and must be handled gently. Usually they are conveyed and washed in large volumes of gently flowing water.

The cans are filled with the segments and are inverted to drain off any excess water. Syrup is then added (approximately 17°B) and the can is vacuum seamed and pasteurised.

6.4.14 Peaches

Peaches for canning should be of a uniform size and symmetrical shape with a good yellow colour and a close tender, but not too soft, texture. In general, the yellow clingstone varieties are most suitable, the most popular being Tuscan, Phillips and the Midsummer type.

The fruit should be picked when it has achieved its greatest size and is still at the firm ripe stage of maturity. Care must be taken when handling the fruit to avoid bruising. Often the fruit is cooled prior to transportation to minimise spoilage. When received at the factory, the fruit is graded by variety and maturity to prevent mixing within the production lines.

The first operation in the canning process is halving and pitting. This can be done by hand or machine. The peaches are cut in half and the pits (or stones) are removed and discarded. Any damaged halves or undersized fruits are separated for use in a sliced pack, or if unsuitable for this, for use in pie filling.

The peach halves are peeled by immersion in a hot caustic soda solution (1–2½%) for 30 to 60 seconds. The caustic soda and loosened skin are removed by passing the peach halves under powerful water sprays. Steam peeling may be used. This loosens the skin, which is then removed by rotating rubber or abrasive rollers.

The peeled peaches are blanched for 1 to 2 minutes in steam or hot water at 80°C to remove the final traces of caustic soda and to inactivate the oxidase enzyme that causes the exposed surfaces of the fruit to turn brown in contact with air.

The peaches are visually inspected, to remove any damaged or blemished halves, and they are then graded. The largest and most perfect halves are the Fancy grade and fruit of smaller sizes and less perfect are known in diminishing order as Choice grade, Standard grade, Second grade and Pie grade. Sliced peaches are prepared from any damaged halves or smaller sized fruit. These are cut into segments and visually inspected to remove any damaged or blemished pieces.

Peach halves are carefully filled into the cans in such a way as to prevent air being trapped in the hollow left by the pit. Hot 30°B syrup is added and the can is closed using steam flow closure.

The cans are pasteurised in hot water or atmospheric steam and are then cooled quickly before being labelled and packed.

6.4.15 Pears

The most popular variety of pear, which lends itself particularly to canning, is the Bartlett (William bon Chretien). It has good texture and flavour with a bright colour and is of a uniform size and shape. The Conference pear is also canned but because of its lower natural acidity the liquor has to be acidified with citric acid.

Pears for canning should be allowed to grow to full size but they should be picked while still green and hard. Ripening is carried out in lug boxes or crates in a well ventilated store. The pears are inspected daily to determine the correct ripeness for processing. A penetrometer may be used to measure the degree of ripeness.

The fruit is mechanically size graded before being peeled, cored and halved using curved knives. After trimming, the halves must be immersed in 1–1.5% brine solution to prevent enzymic browning, which will occur if they are left in contact with the air. Care must be taken to wash all of the salt from the pears before canning.

The pears are filled into the cans, the hot syrup is added and the end is seamed on using steam-flow closure. The cans are pasteurised at 100°C for 12 to 30 minutes depending on the can size. Occasionally, to soften their hard texture sufficiently, the pears need a longer pasteurisation time than is required to achieve commercial sterilisation.

6.4.16 Pineapple

This product is not canned in Europe but it is used as a principal ingredient in fruit salad and fruit cocktail, so it is worthy of consideration in this chapter.

The fruit is harvested when it has reached full maturity but before it has turned soft. On arrival at the factory, it is size-graded to reduce peeling losses. Peeling is carried out on a machine that removes the core and the outer shell and cuts off the shell portion at the ends of a cylinder of flesh. The detached shell is treated by another machine that removes any remaining flesh for use in crushed pineapple or juice.

The cylinders are visually inspected and trimmed to remove any remaining shell or other foreign material. The pineapple cylinders are then sliced to give rings and a further inspection is carried out to remove any misshapen rings and blemishes. Some of the rings are segmented for use as pineapple pieces or chunks.

The fruit is filled into the can, which is vacuum treated to remove any air from within the tissues. Hot syrup is added and the can is seamed and pasteurised prior to being cooled and packed into cases.

6.4.17 Plums and damsons

The best canning varieties are Yellow Egg, Pershore, Victoria and damsons. Some varieties of plums show a tendency to form deposits of gum close to the stone. Usually this cannot be detected on the surface of the fruits. The phenomenon is generally attributed to physiological changes that take place during growth and the enzyme activity associated with the stone. During heat processing, the spots of gum absorb water and swell to form large lumps of jelly that, in extreme cases, may cause the plum to burst and spoil the appearance of the pack. The only way to minimise the problem is by careful selection of raw material.

When the fruit arrives at the factory, the stalks are mechanically removed and the plums are washed. The fruit may be size graded before being inspected and filled into cans. Plain-bodied cans are used for green or yellow varieties while fully lacquered cans are used for the red or purple types.

After filling the syrup is added. It may be possible to use artificial colouring matter in the syrup, but it is advisable to consult current legislation before doing so. The exhausting must be thorough to remove any air, which may cause internal corrosion, and the heat process must be sufficient to inactivate the stone enzymes, which could otherwise accelerate the corrosion process. The pasteurisation time will vary from 10 minutes to 30 minutes depending on the can size and the method of preparation.

6.4.18 Prunes

Canned prunes are an excellent product and are, like fruit salad, able to be canned outside of their normal harvesting season. Dried prunes are produced in the USA and some European countries. Those selected for canning are usually 80 to 90 grade, which is an indication of fruit size, being the number of fruit per 454 g.

Prunes are inspected for broken or defective fruit and for foreign material, after which they are washed and may be blanched. It is possible to soak the prunes overnight, but this will cause much of the natural flavour to be lost in the discarded soak water. The prunes are filled into the cans, allowing space for them to swell as the syrup is absorbed during processing and storage. A UT can will require about 20 prunes, depending on the actual size grade. A low Brix value syrup or fruit juice is added, followed by a thorough exhaust or a vacuum filler to ensure that all the air is removed. This reduces the risk of internal corrosion within the can. The heat process given is longer than that required for sterilisation as the fruit requires softening. The can used should be fully lacquered, usually with a white acrylic lacquer.

6.4.19 Raspberries

This is one of the best-flavoured British fruits and it is popular as a canned pack as it retains its flavour well. Two of the main varieties are Malling Jewel and Glen Clova. They should be picked when ripe but firm and certainly they should never be canned after they have become soft as they will break down completely in the container. Raspberries should be picked into small shallow punnets to avoid unnecessary damage to the fruit.

The raspberries should be canned as quickly as possible after arrival at the factory. It is not advisable to carry fruit overnight as mould grows very quickly, but if the practice has to be followed then the fruit should be chilled to 1–5°C.

The fruit is so fragile that it should not be washed as the action of the water will cause it to break down. The punnets should be shaken gently onto a wide white inspection belt, where inspectors will remove any foreign bodies such as calices, insects, leaf, twig and stones. Unfortunately, some insects such as earwigs or beetles will crawl into the dark recesses of the hole left when the plug has been removed, where they may escape detection.

The raspberries are gently filled into fully lacquered cans, a hot syrup or fruit juice is added and the can is seamed using steam-flow closure. The legislation in some countries may permit the use of artificial colouring matter in this product.

Processing consists of pasteurisation at 100°C to achieve a can centre temperature of 85°C. The processed cans are cooled thoroughly and are gently handled during labelling and casing to prevent damage to the contents.

6.4.20 Rhubarb

The petiole of this plant is regarded as a fruit for food purposes because it is canned in syrup and consumed as a dessert. It forms one of the largest fruit packs in the UK and has the added advantage of maturing very early in the season. Probably the best variety for canning is Champagne, but there are others nearly as good, such as Sutton, Victoria, Albert and Timperley. When the rhubarb is harvested, the leaves are stripped but the bottom end of the stalk is not cut. It is broken from the plant to avoid drying out and subsequent splitting of the stalk.

When the rhubarb arrives at the cannery, the first operation is to cut off the top and base and split any extra-thick sticks lengthwise to avoid over large pieces. The sticks are cut into pieces approximately 25 mm in length, usually by machine. The pieces are then washed in an agitating washer and are inspected. Rhubarb may be lightly blanched to aid filling.

After filling, hot syrup is added to the rhubarb. Artificial colouring matter may be used in the syrup but current legislation should be consulted first. Steam-flow closure may be used when seaming to ensure that all O_2 is excluded from the can.

Because of the high acid content of rhubarb, it may quickly corrode the internal surface of the can. Fully lacquered cans must be used, but even so the shelflife of canned rhubarb is unlikely to be more than a year, more usually 6 to 9 months.

Rhubarb can also be canned as a solid pack. After slicing, washing and inspection, the fruit is blanched at 70°C for 2 to 3 minutes. Artificial colour may be added to the blancher water to aid the colour of the finished pack. When blanched, the rhubarb is hot filled into the cans, seamed and pasteurised.

6.4.21 Strawberries

Unfortunately the colour and flavour of most strawberries deteriorates considerably as the fruit is processed. Both growers and research establishments are continually searching for new varieties that will be more resistant to heat processing. However, in the UK Cambridge Favourite is used almost exclusively for canning where the addition of artificial colour is permitted. There are other red-fleshed varieties that are better for freezing and in canning packs where no artificial colour is added.

In order to obtain a good quality product, strawberries should be handled quickly and carefully. Ideally the fruit should be delivered in shallow trays or punnets. If the fruit arrives in top class condition, it may be possible to avoid washing it. However, experience has shown that washing is usually necessary to remove any adhering soil, straw or leaves, particularly following rain. Washing also helps to remove any residues resulting from spray treatment for fungus or pests.

The fruit should be decalyxed as it is picked, but care must be taken in the factory to remove any residual calices as these are most unpleasant in the finished pack.

Strawberries may be size-graded before packing so that the small fruit are packed into smaller cans, thus avoiding the situation of two or three large berries in a small can. Fully lacquered cans should be used and the fruit should be pasteurised for about 6 to 10 minutes at 100°C. The pasteurising time is determined by the time taken for the temperature at the centre of the can to reach 85°C.

6.5 Bottling

This is not a large industry, although a jar of good-quality fruit looks really appetising on the supermarket shelf. The process is quite simple and is readily carried out in the home.

The choice and preparation of the fruit is similar to that for canning. As the container is transparent, good visual quality is essential. The pack should be free from blemishes and the fruit should be well size-graded. As fruit shrinks on cooking, it should be packed as tightly as possible without damaging it, to avoid giving the impression of an underfilled jar. Filling is assisted if the surface of the jar is wet so that the fruit slips easily. Consequently the jars are used directly after the jar washer where they are inverted and a jet of water is used to remove any possible foreign bodies. After the fruit has been added, the jars are usually completely filled with water, fruit juice or syrup. A small amount of covering liquor is tipped from the jar to give a controlled headspace and then steam is injected immediately before the cup is added to evacuate any remaining air.

The heat treatment may be carried out either in water in a tank heated by steam coils or in a steam/air mixture. In either case, the temperature must be brought up slowly to the processing temperature to avoid thermal shock to the jar, which could result in shattering. Water cooling is not usually employed. If it is used, the hot water must be slowly replaced by cold and the water level must not cover the caps in case any is drawn into the jar by the induced vacuum.

The filled jars should be handled as little as possible for 24 hours after processing as the fruit is very soft and any excessive movement may cause it to break down. Bright light, such as sunlight, should be excluded from store rooms as it will cause the colour of the fruit (particularly red varieties) to deteriorate.

6.6 Freezing

Apart from being a major processing method in its own right, freezing has largely supplanted sulphiting as a means of preserving berries intended for further processing such as jams, conserves and pie fillings. If properly frozen, stored and defrosted, frozen fruit is a high-quality raw material having properties very close to those of fresh fruit.

The freezing process does not destroy microorganisms as does heat sterilisation or pasteurisation, but it essentially retards their growth. Spores may survive the storage process and any damaged microbiological cells may be viable after a period, thus causing consequent spoilage. This means that the fresh fruit must be as free as possible from microorganisms and that scrupulous hygiene is required within the factory to prevent cross-contamination. It is important to handle the material quickly, as bacteria can grow if there is a delay between the stages of processing.

Prior to freezing, the fresh fruit is normally washed to improve the microbiological standard and physical appearance. A fast continuous wash will minimize the leaching of colour and flavour. A close inspection should be carried out to remove any foreign material that might still be adhering to the fruit, together with any blemished pieces.

Fruits in general do not require blanching before freezing. They can be individually quick frozen (IQF) or they can be packed in sugar or syrup or they can be puréed before freezing.

A rapid freezing process is essential to preserve the quality of the fruit. At the freezing point, the ice crystals formed will rupture the cell walls and disrupt the intercellular structure, thus releasing enzymes and substrates that will cause a considerable increase in the rate of deterioration of flavour and colour at this stage of the freezing process. It is important that the fruit is taken below the freezing point as quickly as possible. Rapid freezing yields small ice crystals, whereas a slow freezing process will cause the formation of larger ice crystals that will, in turn, cause greater physical damage to the cell walls. Product damaged in this way will be soft and disintegrated when thawed.

6.6.1 Freezing methods

There are a number of ways of freezing fruit.

Room freezing. This entails placing the product on metal or plastic trays in a cold room at −25°C or colder and allowing it to freeze. If the product is required to be free flowing, it is passed through a breaker while still frozen. This process is often employed for blackcurrants before strigging to prevent excessive damage to the fruit.

Blast freezing. In this process, the product is packed into its final packaging, which may be retail packs or 10 kg boxes, and it is frozen in tunnels with a rapid air velocity at a temperature of −30 to −40°C. This method is not practical if free-flowing product is desired.

Spiral belt freezing. This method is similar to blast freezing in that the product can be in its final packaging. The fruit can also be in pieces, particularly where the pieces are quite large, such as melon balls. It creates less damage than the fluidised bed freezing mentioned below. The fruit passes along a slatted belt which spirals upwards in a chamber where refrigerated air is blown downwards (Figure 6.4).

Fluidised bed freezing. The fruit is fed onto a vibrating perforated bed through which refrigerated air at −30°C is blown from below. The vibrating action of the bed and the turbulence of the air prevent the fruit sticking together and give rise to a free-flowing product. Care must be taken if using this method for soft fruit. Damage to the fruit can be

160 FRUIT PROCESSING

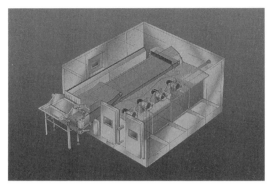

Figure 6.4 A spiral belt air-blast freezer, GYRoCompact (courtesy of Frigoscandia AB).

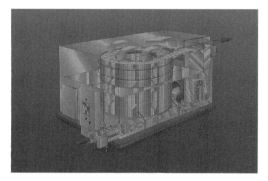

Figure 6.5 A fluidised bed air-blast freezer – FLoFREEZE MX (courtesy of Frigoscandia AB).

minimised by careful adjustment of the loading of the vibrating bed, the rate of vibration and the velocity of the air (Figure 6.5).

Liquid nitrogen freezing. The fruit is fed onto a mesh belt that passes through a tunnel where liquid nitrogen is sprayed onto it. This method is only suitable for small tonnages and it is expensive to operate. It should, therefore, only be used for fruit where a premium price can be obtained or where other methods would cause excessive damage.

6.6.2 Storage

During cold storage, deterioration still takes place, albeit very slowly. It can be shown that for any given storage temperature there is a finite time before there is a detectable deterioration of flavour.

As has been said earlier, a major factor in the success of commercial quick freezing is the small size of the ice crystals formed during the process. However, if the storage temperature is not correct or there are fluctuations of temperature, the ice crystals will combine during storage to large crystals. This obviously defeats the object of fast freezing. Therefore, the control of temperature during storage and distribution is as important to the quality of the food as is the original process.

It is generally considered that below $-12°C$ microbial growth ceases or is, at worst, extremely slow. Again, it is essential to maintain the correct storage temperature; the product must not be allowed to thaw as this will allow any surviving microbes to grow. Thawing followed by re-freezing will cause the ice crystals to grow larger and causing rupture of the cell structure and a product with a poor texture.

Immediately after freezing, the products are placed into cold stores where the temperature is maintained at between $-23°C$ and $-30°C$. The products are then distributed in insulated containers that are equipped with their own refrigeration units. Where refrigeration units are not available, the low temperature can be maintained by using 'dry ice', i.e. solid CO_2, on racks in the container. This is only satisfactory for journeys of relatively short duration. In order to ensure that the product temperature does not rise on transfer from the cold store, it is advisable to load the containers from the cold store using 'port doors', where the container is backed flush to the door to form a seal, thus preventing warm ambient air coming into contact with the product.

On arrival at the retailer, the products should be stored in a small cold store or be placed directly into the display cabinet where the temperature is maintained at $-18°C$. Care must be taken not to overload these cabinets, as thawing can occur if the product is above the marked load line.

6.6.3 Packaging

In addition to the requirements common for all food containers, such as attractive design, consumer convenience, construction from non-toxic materials and economic costs, frozen food packaging must be able to withstand the low temperatures required by the process and it must prevent the product suffering from freezer burn during storage. The pack must stay in one piece even when it becomes damp as the product is thawing. Strength is also required, as the product is handled in the retail display cabinet.

Tests show that the rate of weight loss from a pack during frozen storage is governed by the permeability characteristics of the packaging material. A pack with a good moisture barrier but not completely filled with product is at risk of ice separation caused by sublimation from the product if it is subjected to fluctuating temperatures.

Another requirement of the packaging is that it should be capable of being handled by high-speed machinery. Much use is made in the frozen food industry of form, fill and seal packaging.

Packaging can take the form of waxed cartons, direct film wraps and bags or overwrapped trays. All of these can be used for fruit. Cartons can be used for fragile fruits such as raspberries, bags for more robust products such as rhubarb or apple pieces and the overwrapped trays may be used for precooked fruit pies.

6.7 Aseptic packaging

Aseptic packaging is another form of packaging heat-processed fruit products. In contrast with canning and bottling where the product is filled into containers, sealed, sterilised and cooled, aseptic technology requires the product to be sterilised or pasteurised using some form of heating and then filled under sterile conditions into sterilised packaging and finally sealed. The process is not used very extensively for particulate fruits. It is reserved more for juices and purées (see chapter 4).

The chemical and physical changes are not so severe with fruit products as with others. Many of them are acid and, therefore, relatively low temperatures are required for sterilisation. Traditionally many of the fruit products have been hot packed.

6.7.1 Sterilisation/pasteurisation

This is carried out either by direct heating with steam or by indirect heating using some form of heat exchanger (e.g. tubular, plate or scraped surface).

The advantage of the aseptic process is that it is necessary to use a high temperature (approximately 150°C) for only a few seconds, which ensures that more of the flavour and the nutrients are retained compared with the lower temperature and longer time sterilisation demanded by the canning process. The short time is made possible because the product is heated without a container to restrict the heat transfer by holding the contents in bulk. In aseptic processing, the product is heated quickly in a thin film. Certain changes do occur when processing these products at elevated temperatures. Oxygen is removed during the process, which helps in the colour retention of certain products. However, in other products, such as bananas, the browning or discoloration caused by oxidation is extremely serious. In these cases, the processing system should be arranged to include efficient de-aeration as soon as possible after peeling.

Over the years, the processing of particulate fruits has been achieved using a number of methods. Particles up to 5 mm in size can be processed in a tubular heat exchanger, whereas larger particles, up to 15 mm, can be processed in a scraped surface heat exchanger with a 15–18 mm gap between the rotor and the heating surface.

There are special types of heat exchanger that incorporate product flow separation, thus permitting different residence times for particles and liquids. A typical device is Rotahold, which consists of a cylindrical vessel fitted with infeed and discharge ports. Inside, a set of fork blades are mounted on a central shaft that can be rotated at an adjustable speed to suit the residence time. The fork blades are separated so that the particles are retained and the liquid passes through the particles. A set of stationary blades between the infeed and outlet prevents the particles short-cutting the system.

Particles and liquids can be processed separately; the particles being batch processed in a jacketed vessel that rotates to ensure even heating. The hot liquid is added from another heating vessel and the whole is then filled into the containers.

Aseptic processing is also used for purées, such as fruit pie fillings, in a number of ways:

- steam injection, where live steam is injected directly into the purée
- a tubular heat exchanger, where the product passes through a stainless steel tube that is heated on the outside by steam. The tube may be coiled for greater efficiency
- a plate heat exchanger, which allows the product to flow through alternate channels in a pack of embossed plates with the heating medium flowing through the other channels
- a scraped surface heat exchanger, in which the product is moved along a heated tube by a rotating scroll or blades.

Good hygiene is of paramount importance in aseptic processing. The whole technology is based on achieving sterility before the product is filled into the packaging. This means that there must be extensive hygiene schedules and cleaning instructions and these must be carried out assiduously.

6.7.2 Packaging

There is a wide variety of packaging available for aseptic products including:

- rigid containers, such as metal cans and glass bottles and jars
- semi-rigid containers, such as plastic cans, bottles and tubs, aluminium trays and paper laminated cartons
- non-rigid containers, such as plastic pouches and bags.

The choice of packaging is determined by the food product, its expected shelflife, marketing appeal and cost. There are other criteria affecting the choice, including tamper evident to prevent product contamination, physical protection of the product and also the need to provide the consumer with information about the product.

Aseptic products have the advantage shared by canned and bottled products that they can be stored at ambient temperatures. In these days of high-energy costs, this must be an advantage.

Further reading

Anon (1984). *Handbook for the Fruit Processing Industry*. A/S Kobenhavns Pektinfabrik, Denmark.
British Fruit and Vegetable Canners' Association. *Code of Practice on Canned Fruit and Vegetables*.
Campden and Chorleywood Food Research Association (1980). *Revised Technical Bulletin*, No 4, Chipping Campden, Glos.
Carnaud Metalbox plc. *Double Seam Manual*.
Cherry Burrell Application Engineering Department (1982). *Aseptic Processing*. Cherry Burrell Corporation, Cedar Rapids, Iowa, USA.
Cruess, W.V. (1958). *Commercial Fruit and Vegetable Products*, 4th Edition, McGraw-Hill Book Company Inc., New York.
Hanson, L.P. (1976). *Commercial Processing of Fruits*. Noyes Data Corporation, New Jersey, USA.
Holdsworth, S.D. (1992). *Aseptic Processing and Packing of Food Products*. Elsevier, London.
Holdsworth, S.D. (1983). *The Preservation of Fruit and Vegetable Food Products*. Macmillan Press, London.
Nagy, S., Shaw, P.E. and Veldius, M.K. (1977). *Citrus Science and Technology*. AVI, Westport, USA.
Ranken, M.D. (ed.) (1988). *Food Industries Manual*. Blackie Academic and Professional, Glasgow.
Rees, J.A.G. and Bettison, J. (eds) (1991). *Processing and Packaging of Heat Preserved Foods*. Blackie Academic and Professional, Glasgow.
Shapton, D.A. and Shapton, N.F. (1991). *Principles and Practice for the Safe Processing of Foods*. Butterworth–Heinemann, Oxford.
Tressler, D.K., van Ardsel, W.B. and Copley, M.J. (1968). *The Freezing Preservation of Foods*. Vol. 1 – Refrigeration and Equipment. AVI, Westport, USA.
Various (1982). Papers presented at *A Seminar on Aseptic Processing and Packaging*. 18–19 October 1982. Centre for Continuing Education, University of Aukland, New Zealand.
Willhoft, E.M.A. (1993). *Aseptic Processing and Packaging of Particulate Foods*. Blackie Academic and Professional, Glasgow.

7 The manufacture of preserves, flavourings and dried fruits
R.W. BROOMFIELD

7.1 Preserves

The brief for this chapter was that it should be 'state of the art'. For those involved in the manufacture of preserves, it often appears that there is more 'art' in the job than science. There is a mystique in 'stirring the pot', which at the time often seems difficult to explain: experience is a very valuable commodity. Nevertheless, when one, in the relative peace away from the manufacturing floor, tries to seek an explanation to a problem recently encountered (and problems there will always be), a scientific reason always comes to the fore. Unfortunately, however, there are two, three or more possible explanations for the same problem and the sorting out of the interlinked relationships of these can cause sleepless nights.

It is now over forty years since the book that became the standard work on commercial jam-making first appeared; the first edition of *Jam Manufacture* by G. H. Rausch was published in 1950. Much of the information it contained is most relevant today, although many of the suggestions are dated when compared with the requirements of today, particularly those relating to food hygiene. Governing jams, jelly jams, marmalade, lemon curd and mincemeat at that time was the 1953 *Preserves Order* (Statutory Regulation and Order 1953 No. 691, as amended), which had replaced earlier (1944) legislation. This has, in turn, been replaced in the UK by the *Jam and Similar Products Regulations* (1981) (Statutory Instrument 1981 No. 1063) which enacted the EC Directive *79/693/EEC*, and which has itself been amended by the EC Directive *88/593/EEC*.

Apart from differences in prescribed minimum fruit contents (and the introduction of several additional categories of product), a major difference between the 1953 and the 1981 regulations is the prescribed content for total soluble solids. For 'Jam', the 1981 Regulations prescribe a minimum figure of 60% soluble solids (as measured by refractometer) compared with 65% (for an hermetically sealed container) and 68.5% (non-hermetically sealed) in the 1953 order.

What is jam? A summary description from the 1981 Regulations defines it as 'a mixture of fruit . . . and sweetening agents brought to a suitable gelled consistency, with or without other permitted ingredients' (or words to that effect).

The traditional understanding of 'jam' was that of a self-preserved cooked mixture of fruit and sugar (honey often qualified as a sugar). The higher figure of 68.5% soluble solids did give a degree of self-preservation which is not achieved by the 1981 limit of 60%. The degree of preservation is related to the water activity of the product, but there are other factors affecting spoilage. In the light of experience, two jams can be prepared and stored similarly and have identical analytical characteristics in relation to soluble solids content, pH and titratable acidity, and yet the one made from apricots will spoil more quickly than the one made from blackcurrants. Research into this phenomenon could prove useful!

The 1981 UK legislation prescribes other products such as 'reduced sugar' preserves. These, by themselves, are not self-preserving but need the presence of a permitted preservative to prevent spoilage.

For the labelling requirements of preserve products, one should refer to the appropriate legislation.

7.1.1 Ingredients

The ingredients in preserve manufacture are fruit, sweetening agents and any other permitted ingredient that may be required. The last group includes gelling agents, acids, buffer salts and colours, although other substances such as spiritous liquor may be used, particularly in speciality products. In fruit curds, a source of fat (normally either butter or margarine) is a mandatory ingredient; fruit curds also contain egg. That traditional British commodity, mincemeat, comprises, at its most basic, sugar, apples, vine fruits, citrus peel, a source of fat, spices and acetic acid.

7.1.2 Fruit for jam manufacture

'You cannot make good jam without good fruit' is a maxim often overlooked, for it has frequently been the case that overripe or underripe fruit has been used for processing. Underripe fruit rarely has the flavour characteristics and the developed colour of fully ripe fruit and the natural pectin present at that stage is often unusable to the jam maker; pectin becomes more soluble, and so more available, as the ripeness of the fruit increases. Overripe fruit often has poor flavour and is subject to microbiological spoilage. In addition, enzymatic action will have broken down the pectin, leading to disintegration of the structure of the fruit.

Very few large commercial concerns now produce jams and marmalades from fresh fruit during the harvesting season; most use fruit that has been preserved one way or another. Three types of preservation are used: freezing, canning and sulphiting. The first two of these are dealt with in the previous chapter and will be discussed very briefly here.

Freezing is a good method for maintaining fruit quality although, to the jam-maker, the structural changes that occur in the fruit through ice crystal formation may adversely affect the texture of the finished preserve.

In the UK, canned fruits tends to be limited to imported fruits such as apricots, peaches and pineapples. These fruits are often packed to the purchaser's requirements. A development of this is the use of aseptic packing, either truly aseptic, where the fruit is sterilised by heat and then cooled and filled aseptically into sterile containers, or hot filling into semi-bulk containers, such as internally laquered drums of 200 litre capacity, that are sealed and crash cooled by immersion in cold water.

Chemical preservation of fruit intended for jam manufacture using sulphur dioxide is widely practised and is easy and cheap to carry out. A sulphur dioxide concentration of 2000 to 3000 mg/kg of fruit is achieved by using a 6% aqueous solution or the gas may be added directly to the fruit. The use of salts of sulphurous acid is also common but can result in problems later on in the manufacture of the jam because of the presence of metallic (e.g. sodium) ions, which affects the pH of the finished jam. Storage of sulphite-preserved fruit may be in polythene lined drums or, where whole fruits are not present, in bulk storage tanks. The presence of whole fruits in bulk storage tanks can lead to poor fruit piece distribution as a result of floating of the particles.

There are legal constraints on the use of fruit preserved with sulphur dioxide: it may not be used, for example, in Extra Jams or Extra Jellies. The limitations on its use are mainly concerned with its removal during subsequent processing. For example, it is easier to boil off sulphur dioxide using open pans than when processing under vacuum. Atmospheric pollution must also be considered because sulphur dioxide is blamed for acid rain. A factory producing 1000 tonnes of jam from sulphited fruit pulp by open pan boiling, with discharge of the evaporated steam with its entrained sulphur dioxide into the atmosphere, can result in the release of up to 1 tonne of the gas.

Before many fruits can be used in jam manufacture, a pre-processing operation is often necessary, for example, peeling, slicing, dicing and/or cooking. Many of these operations can be carried out at the jam-maker's factory; more often these days, the fruit is bought in as a ready-to-use ingredient. Some examples of fruit processing are discussed later in this chapter.

Types and varieties of fruit. Almost any type and varieties of fruit can be used in the manufacture of preserves but, inevitably, some fruits will find more favour than others. The manufacture of jam has been in decline for many years, and, in consequence, many of the fruits used in earlier times are no longer commercially available although they may sometimes still be obtained by the maker of speciality products. A typical example is the strawberry, where simple economies has resulted in an almost total loss of British grown fruit for use by the larger British manufacturer.

Apples are used in the manufacture of many types of preserve, especially bakery preserves and in mincemeat. For use in the former, the fruit is cooked and sieved to produce a purée, while in the latter it may be used as pre-cooked purée or in an uncooked macerated or diced form. The most popular apple in use today is the Bramley Seedling.

Blackcurrants can be used as sulphited blackcurrant pulp or in whole fruit form (rather than sieved). The latter should be precooked before inclusion in the jam-making process to prevent shrivelling of the berries and a consequent toughening of texture.

Many fruits have lost popularity over the years, or may not ever have been as popular as the main-line varieties of today, for example, the gooseberry, damson and plum. Earlier writers have promoted the use of underripe plums and it is felt that this may have had a deleterious effect on the sales of plum jam for while the use of sulphited fruit enables manufacture from a basically cheap raw material to be carried all year round, there are differences in flavour from similar jams made from sulphited or fresh/frozen fruit. These comments apply to all fruit varieties and the manufacturer has to choose between the cost and flavour constraints when bringing a product on to the market. The fickle nature of the consumer must also be considered!

Citrus fruits and, in particular the bitter orange, are used in the manufacture of marmalade; indeed the term marmalade is reserved for products made with citrus fruits. The oranges (or other fruit) must be preprocessed before the marmalade can be made.

7.1.3 Other ingredients

Sweetening agents. The most commonly used sweetening agent is white sugar (sucrose), either as the dry ingredient or as an aqueous solution or syrup. Glucose syrup is also in general use in preserves, although the compositional requirements of the end-product must be considered because it often contains very small quantities of sulphur dioxide. Other commercially available syrups (e.g. high fructose syrup or invert syrup) can be used, as can honey. These syrups must be used carefully because the invert sugar content of the finished product can affect the quality of the set as well as the potential for crystallisation, particularly in jams of higher total soluble solids content (such as some bakery jams).

Acids and buffers. 'Fruit' acids, commonly citric or malic acids, are often added to adjust the flavour profile and to achieve the optimum pH for setting. Buffer salts such as trisodium citrate are also used to achieve pH. Lemon juice may be substituted for citric acid.

Fats (for curds and mincemeat). Butter or margarine is normally used in fruit curds, while beef suet or its vegetable equivalent, normally in shredded form, may be used in mincemeat.

Vine fruits (for mincemeat). Details are to be found later in the chapter.

Citrus peel (for mincemeat). The peel is separated from the centre of the citrus fruit and often held in brine prior to dicing, tenderising and soaking in sugar syrup. This is drained prior to use. Normally, lemon and sweet orange peels are used, although it is not unknown to use bitter orange peel in mincemeat.

Gelling agents. Pectin is the gelling agent that occurs naturally in fruit and it governs the setting characteristics of the product. The pectic substances that occur naturally in the middle lamella of plant tissues change with the changing maturity of the fruit. They can be regarded as the 'glue' that holds the plant cells together. As the fruit ripens, an insoluble pro-pectin is converted into souble pectin. The pectin molecule is composed of long chains of partially methoxylated polygalacturonic acid (Figure 7.1). In riper fruit, enzyme action will break the chains into shorter lengths, producing pectin with inferior setting characteristics. The quantity and the effectiveness of pectin also vary with the variety of fruit. Citrus fruits and apples contain more pectin than do cherries, for example. It is prudent, therefore, for the jam-maker to estimate the useful pectin content of his fruit before considering large-scale production.

Because of the need to produce a consistent product in the face of natural variations, pectin often has to be added to the product. This is essential when depectined fruit juices are used in the production of jelly jams! Commercially, pectin is extracted from both apples and citrus fruit, or rather, more normally, from apple residue after the extraction of the juice, and from citrus peel. From UK production, apple pectin is sold as a liquid while citrus pectin is sold as a powder. Both sources will normally have been standardised before sale.

Naturally occurring pectin is termed 'high methoxyl', with a sufficiently high percentage of its hydroxyl groups methylated for it to fall into the 'fast set' category. This will form a gel with high sugar content solutions

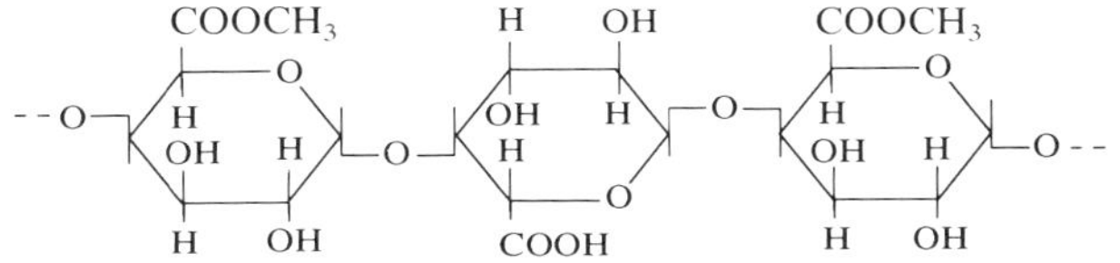

Figure 7.1 The structure of part of the pectin molecule.

(refractometer solids between 60 and 70%) and with pH values between 2.8 and 3.5. The manufacturers of pectin chemically modify the pectin molecule in order to affect the speed of setting of the gel. A 'slow set' high-methoxyl pectin will have between 60 and 68% of its groups methylated while a 'fast set' pectin has between 68 and 75%.

When a high-methoxyl pectin is dissolved in water, it forms a slightly acidic solution, and there is a tendency for the carboxylic acid groups to dissociate; this is controlled by adjusting the pH. The addition of sugar has a dehydrating effect on the pectin and will reduce its solubility. To form the gel, cross-linkages are formed when these two effects are combined.

The set, its quality and speed of production are governed by the degree of methylation and the quantity of the pectin and also by the sugar content (and the types of sugar present), the pH and the temperature of the solution. A fast-set high-methoxy pectin will have a higher setting temperature than a similar gel made using a slow-set high-methoxyl pectin. For the jam-maker, it is sometimes desirable to use a slow-setting pectin; for example, where the flotation of particles of fruit to the surface of the jam is not a problem. Normally, however, a compromise is needed, and a balance between the variables achieved.

Modifications to the pectin molecule to reduce the number of methoxyl groups while maintaining the designation 'high methoxyl' have been mentioned. If further modification takes place with the removal of more methoxyl groups, the pectin becomes 'low methoxyl', and will produce gels at lower sugar contents. Further chemical modifications can produce 'amidated' pectins, where up to 20% of the methoxyl groups have been converted to amide groups. Both ordinary low-methoxyl and amidated pectins behave similarly and are dependent on the presence of calcium for gel formation, with the gel strength being less dependent on sugar content. Indeed, some of these pectins can form gels with sugar contents of below 20%. These are important in the manufacture of 'Reduced Sugar Jams' where the low sugar content prevents gel formation using the pectin naturally present in the fruit.

Other gelling agents, such as alginates or others prescribed in the Regulations, may be used, often in collaboration with other stabilisers in the manufacture of reduced sugar jams. It is recommended that the intending user experiments with various combinations until a satisfactory mix is arrived at.

7.1.4 *Product types and recipes*

Compositional definitions from the 1981 UK Regulations include Extra Jam, Jam, Reduced Sugar Jam, Extra Jelly, Jelly, UK Standard Jelly, Marmalade, Reduced Sugar Marmalade, Mincemeat, Fruit Curds and Fruit Flavour Curds and Sweetened Chestnut Purée. There are two further categories that are deserving of mention: No-Added-Sugar Fruit Spreads and Diabetic Jam. Mention will also be made of products prepared specially for manufacturers: Bakery Jams. The Regulations also include many more definitions, including those of the permitted sweetening agents and other authorised ingredients.

Sweetened Chestnut Purée is a mixture of sweetening agents and puréed chestnuts (minimum 38 g/100 g) with minimum refractometer solids of 60%.

Jams and Jelly Jams are similar except that the former products are made with whole fruit, fruit pulp or fruit purée, while the jelly jams are made using fruit juice. There are two qualities of Jam: Jam and Extra Jam. Three qualities of Jelly Jam are specified: Jelly, Extra Jelly and UK Standard Jelly. Each will be dealt with separately.

The general fruit content for Jam is 35 g of edible fruit per 100 g of product, and the product can be made from fruit, fruit pulp or fruit purée (or a mixture of these). There are some exceptions to this general limit, most notably blackcurrant (25 g per 100 g) and ginger (15 g per 100 g). The fruit used for this category of product may have been preserved with sulphur dioxide, and in recipe calculations, the added preservative (and any other ingredients such as water) used in the manufacture of the pulp must be considered. The term pulp here is a jam-maker's term for a ready-to-use ingredient and should not be confused with the legal definition!

For Extra Jam, the general minimum fruit content is 45 g of edible fruit per 100 g, with similar exceptions to those for Jam: blackcurrant at 35 g per 100 g and ginger at 25 g per 100 g. Extra jams cannot be made from purée, and the fruit must not have been preserved with sulphur dioxide. In addition, Extra Jams cannot be made with mixtures that include apples, pears, clingstone plums and several other fruits. It must be pointed out that this does not prevent the manufacturer from making products using mixtures where the fruit content is above the legal minimum for Extra Jam, but the product cannot be called an Extra Jam.

The recipe for a product is a very personal thing, and its make-up is often jealously guarded, for commercial viability often depends on its contents being unknown to others. Below, therefore, are the basic ideas of formulation, which may prove useful to the reader.

When formulating recipes, it is normal to specify a standard theoretical output and relate ingredient quantities to this. It is necessary to know the actual fruit contents (both 'total', for the requirements of the labelling legislation and 'edible', for the recipe itself). Certain analytical details of the ingredients also need to be known.

It may be necessary to add a quantity of a gelling agent to the product: pectin is the most common in Jams and Extra Jams, and should be added as a solution.

Sugar (sucrose) is the predominant sweetening agent in most Jams and

Extra Jams. During the boiling operation, some of the sucrose will be inverted to form glucose and fructose (invert sugar). One molecule of sucrose ($C_{12}H_{22}O_{11}$) combines with one molecule of water (H_2O) to form one molecule of glucose and one of fructose (both $C_6H_{12}O_6$). Thus 1.000 kg sucrose will produce 1.052 kg invert sugar. For the purpose of the illustrations below, it is assumed that 25% of the sucrose is converted into invert sugar during processing; in which case, 0.987 kg dry sugar will produce 1.000 kg of the mixture of sucrose and invert sugar. This supposition is not atypical of normal processing.

An example of formulation of a recipe. To prepare a raspberry jam that is to contain the UK minimum fruit content of 35 edible fruit per 100 g of product would use that amount of whole raspberries. For a seedless raspberry jam, the contribution of the raspberry seeds (approximately 4% by weight) would need to be considered and 36.4 g of raspberries per 100 g product would be needed. The exact figure is determined empirically.

The label declarations on the raspberry jam would be 'prepared with 35 g of fruit per 100 g', and it would have 'total sugar content of 66 g per 100 g'. If the particular jam is to be made from sulphited pulp, which is specified as having a total (edible) fruit content of 92% by weight, the 100 g of product would contain 38.04 g (35 ÷ 92%) of the sulphited pulp. The total soluble solids of the pulp, as measured by refractometer, is found to be 8% and so the 38.04 g of pulp contains 38.04 × 8% = 3.4 g of sugar solids.

By experiment, the pectin requirement is set at 10.0 g pectin solution per 100 g of product. This pectin solution contains 10.0% refractometer solids and, therefore, contributes 1.00 g of solids. The final refractometer solids of the jam are to be 66%; so the sugars to be added are 66 minus 3.04 (from the fruit) minus 1.00 (from the pectin), i.e. 61.96 g per 100 g of product. This is added as dry sugar (sucrose), and it is assumed that 25% of the sugar is inverted during processing. The amount of dry sugar which has to be added is, therefore, 61.96 × 0.987 = 61.16 g per 100 g (i.e. the 'sugar equivalent' is 101.32% of the sucrose content and reflects the refractometer readings).

Finally, by experiment, it is found necessary to add buffer salts (citric acid and, perhaps, sodium citrate) in small quantities to produce satisfactory conditions for setting (see also the section on pectin below). In addition, the use of added colour may be needed, especially where sulphited red fruits are used, as the whole of the colour does not return during the boiling that drives off the sulphur dioxide.

These calculations may be (perhaps more simply) represented using a proforma recipe sheet as shown in Table 7.1.

The sugar could be added as sucrose syrup (or 'sugar solution', as defined in the UKs Specified Sugar Products Regulations 1976 (S.I. 1976

Table 7.1 A proforma recipe sheet for Jam

Sugar equivalent (kg)	Ingredient	Quantity (kg)
3.04	Raspberry pulp (8.0% TSS)	38.04
61.96	Sugar	61.16
1.0	Pectin solution (10.0% TSS)	10.00
	Colour	
	Citric acid	} If required
	Sodium citrate	
Refractometric solids, total 66.0		Boil to output 100.00

Table 7.2 Proforma recipe sheet using sucrose syrup

Sugar equivalent (kg)	Ingredient	Quantity (kg)
3.04	Raspberry pulp (8.0% TSS)	38.04
61.96	Sugar Syrup (67°B)	91.28
1.0	Pectin solution (10.0% TSS)	10.00
	Colour	
	Citric acid	} If required
	Sodium citrate	
Refractometric solids, total 66.0		Boil to output 100.00

No. 509), in which case, due account would have to be taken of its concentration. Such syrups are often supplied at 67% sucrose (67°B by refractometer), and if this were used, the recipe sheet would appear as in Table 7.2. It is also quite normal, technically acceptable and often commercially beneficial for up to 30% of the added sugar to be replaced by commercial glucose syrup. Such syrups may be slightly less sweet than sugar or invert syrup, and it is often thought that a decrease in sweetness is desirable in today's preserves.

If 25% of the originally calculated added sugar were to be replaced with a glucose syrup having 80% total soluble solids, the recipe sheet would now appear as in Table 7.3.

This is only one example; the proportion of sugar solution and glucose syrup can be varied as desired.

Small errors will be apparent in these calculations; no account of the contribution of acid and buffer to the refractometer solids has been taken and the degree of inversion has been assumed to be 25%; in practice, this may not be the case.

Once a formulation has been decided upon, it is advisable to carry out

Table 7.3 Proforma recipe sheet using glucose syrup

Sugar equivalent (kg)	Ingredient	Quantity (kg)
3.04	Raspberry pulp (8.0% TSS)	38.04
46.48	Sugar solution (67°B)	68.46
15.48	Glucose syrup (80°B)	19.35
1.0	Pectin solution (10.0% TSS)	10.00
	Colour	
	Citric acid	} If required
	Sodium citrate	
Refractometric solids, total 66.0		Boil to output 100.00

Table 7.4 Recipe for Raspberry Extra Jam

Sugar equivalent (kg)	Ingredient	Quantity (kg)
3.60	Raspberries, fresh or frozen (8.0% TSS)	45.0
62.00	Sugar	61.19
0.4	Pectin solution (10.0% TSS)	4.00
	Citric acid	
	Sodium citrate	} If required
Refractometric solids, total 66.0		Boil to output 100.00

experimental boils to ascertain the exact needs in terms of added pectin, acid and buffer.

At first sight, the first of the recipes shown above may appear to be the most economic (neglecting any differences there might be in the prices of sugar and glucose syrup) as it exhibits the least amount of required evaporation. This is not necessarily the case because of the possibility of the precipitation of pectin from solution if insufficient water is present. Experience has shown that a boil size of 120 g minimum per 100 g output is required (for normal jams) if this phenomenom is to be avoided, for once the pectin has lost its solubility, the boil seldom can be satisfactorily reclaimed and properly set jam produced.

The UK (and EU) requirements for an Extra Jam are prescribed, in that the minimum fruit content and the type of fruit to be used are defined. The formulation for a Raspberry Extra Jam is calculated similarly to the above. To make a product at the statutory minimum fruit content, and with a sugar content of 66 g per 100 g, the recipe might read as in Table 7.4.

The basic difference between Jams and Extra Jams is the additional fruit content. This is an oversimplification, because of the differences relating to residual sulphur dioxide and the use of added colours, for example. To produce a product that is acceptable legally, one must refer to and comply with the legislation of the country in question.

Jelly jams, in all their guises, are similar to normal jams but the source of fruit is juice rather than whole fruit, fruit pulp or fruit purée. The recipes can be worked out in the same manner as those outlined above. UK and EU legislation provides for three categories of jelly jams: Jelly, Extra Jelly and UK Standard Jelly. A Blackberry Jelly must contain a minimum of 35 g of blackberry juice per 100 g of product, a Blackberry Extra Jelly a minimum of 45 g while the UK Standard Blackberry Jelly harks back to older legislation and is required to contain the juice of at least 35 g of blackberries per 100 g of product.

The juices used for jelly manufacture are normally clarified and are often treated with a pectinase enzyme to destroy the natural pectin. In this case, it will obviously be necessary to add pectin in order for the product to set!

Marmalade is also compositionally prescribed in UK and EU law (as well as in that of many other countries). It is a jam that is manufactured exclusively from citrus fruit, and the normal fruit to be used is the Bitter (or Seville) Orange. The legislation requires marmalade to contain a minimum of 20 g of citrus fruit per 100 g of product. The fruit is further defined as pulp, purée, juice, aqueous extract or any combination thereof with the proviso that at least 7.5 g is derived from the centre of the fruit. Although the bitter orange is the fruit normally used in the UK product, other citrus fruits such as grapefruits and sweet oranges are finding favour.

Jelly marmalades are among the most popular of marmalades and are made with juice extracted from the fruit and which is normally clarified before use. It is often concentrated in the country of origin and exported in this condition; the degree of concentration must be known exactly if the fruit content of the finished product is to be correctly assessed. The marmalade may or may not contain thin-cut shreds of peel, which are usually about 25 mm long, 1.5 mm wide and 1.5 mm thick, i.e. without albedo. They are almost invariably tenderised and soaked in syrup before use in order to obtain a even distribution of shreds throughout the marmalade.

Marmalades containing medium or fine-cut peel are normally made with a purée of the centre of the fruit and shreds of peel. The mechanical cutters which separate the peel from the centre are set to provide peel with the maximum amount of albedo. The shreds have a maximum length and a width between 1.5 and 3.0 mm. Such marmalades normally have the peel pieces tenderised (by cooking in water or steam) and then mixed back with the pre-cooked and sieved centre purée before the mixture is used to make marmalade. Pulps made up of peel and centre purée are available commercially having been prepared in the country of origin and packed either aseptically or with the preservative sulphur dioxide. Whether the pulp is made in the factory where the marmalade is to be prepared or elsewhere, it is essential to ascertain its exact fruit content in order to establish the fruit content of the finished marmalade.

Marmalades with thick or coarse cut peel contain much larger pieces of peel – sometimes up to 8 mm wide and 50 mm long – which are, after tenderising and prior to marmalade manufacture, sometimes impregnated with some of the recipe requirement of sugar to increase their density (in a way similar to the peel used for jelly marmalade. This helps to achieve a good distribution of shreds in the finished marmalade. It is generally recognised as producing a marmalade with better flavour (especially when fresh fruit is being used) and helps to achieve a good 'clean' set (with a minimum of syneresis). The method is, however, considered to be too time consuming these days, being both inefficient and expensive.

Formulations for marmalades are worked out in an identifical manner to the examples shown earlier for raspberry jam, but it cannot be emphasised too much that in order to produce a consistent product, one must know precisely the way in which the raw material is treated and its composition. A more detailed discussion of the preparation of citrus fruit for marmalade manufacture is found later in this chapter.

The basic principles outlined above apply equally well when applied to Reduced Sugar Jams and Marmalades. It must always be remembered that, in these products, preservatives are permitted (as prescribed in the legislation) because the products themselves are not self-preserving. It must also be remembered that if a low-methoxyl or amidated pectin is used, the setting characteristics of the product are dependent on its calcium content. Some of these reduced sugar products are sold as 'made especially for diabetics' because their formulations include only half the amount of added sugar present in ordinary jam. This follows a requirement in the UK Food Labelling legislation. Traditional Diabetic Jam, however, uses sorbitol as a replacement for the added sugar, as sorbitol, a polyhydric alcohol, does not affect the metabolism of the diabetic. The same manufacturing methods are used, but care must be taken to ensure that the products are not contaminated by sugar. Many commercially available pectin powders are standardised using sugar and so the maker of Diabetic Jam should ensure that his ingredients are not so treated.

7.1.5 Methods of manufacture

General. Figure 7.2 is a flow diagram of the basic processes in preserve manufacture. This can be extended to include all the different stages in the process and in such a form could form the basis of a hazard analysis for the product. The addition of control points (which may or may not be critical control points) could be most useful in the production of a full HACCP

PRESERVES, FLAVOURINGS AND DRIED FRUIT 177

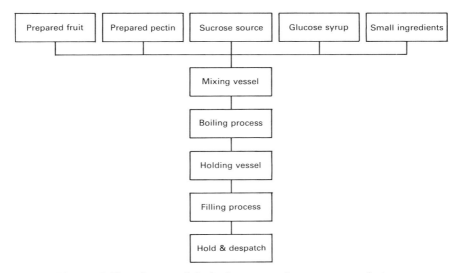

Figure 7.2 Flow diagram of the basic processes in preserve manufacture.

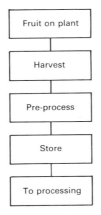

Figure 7.3 Preprocessing stages in preserve manufacture.

(hazard analysis critical contol points) assessment. Control points have not been added into this basic flow chart.

Fruit preparation. Almost all fruit will require preprocessing in one way or another before it is in suitable condition for incorporation into the mix to be used for the preserve (Figure 7.3). This particularly applies to citrus fruit. The preprocessing of other fruits will often involve trimming and perhaps tenderising to facilitate subsequent osmosis and to give a satisfactory mouthfeel to the preserve. Some fruits (e.g. apricots) may be canned as solid pack to extend the period that the fruit is available and this also achieves the tenderising of the fruit!

All fruit should be properly sorted before use to exclude foreign bodies or other unacceptable material including that from the source plant.

The processing of citrus fruit is, of necessity, more complicated. The peel is separated from the centre of the fruit, and the two parts are treated separately, often being recombined in natural ratio for use in marmalade manufacture. Sometimes, however, it is necessary for the two parts to be kept separate, only to be recombined when the marmalade is being produced. Sometimes these ingredients may not be recombined in a

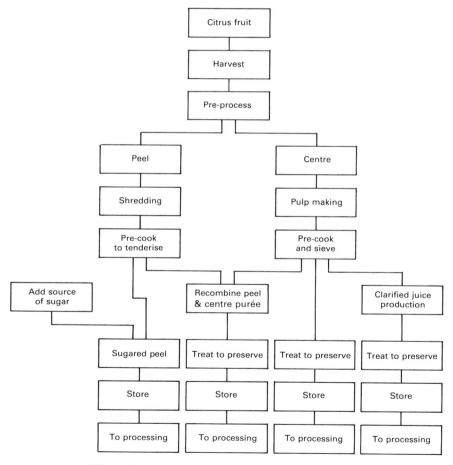

Figure 7.4 A flow diagram for citrus fruit preparation.

natural ratio, where, for example, a high shred content is required. Shredless marmalades are also manufactured. Figure 7.4 shows a flow chart for citrus fruit preparation. As with the general chart, this can easily be extended and used in a HACCP presentation.

Boiling and filling. For normal jams and marmalades, some evaporation is needed to move from the mixing bowl stage to the finished product. Reduced sugar products, however, can be formulated so that evaporation is not necessary. For some products, a two-stage process may be preferred if certain potential problems (such as the floating of fruit particles or pieces in the finished product) are to be avoided.

The boiling of jams is outlined below. Because this process can vary greatly, not only with the method of boiling used but also with the different types of fruit and even with different batches of the same fruit, acquired experience is invaluable.

Traditionally, jam boiling was carried out using open (atmospheric) copper boiling pans, but these days pans are now usually stainless steel (Figure 7.5). The pans are hemispherical, with a large lip extension to assist in the prevention of boilover. The hemispherical part of the pan is steam jacketed, with high-pressure steam providing the heat source. Extra heating may be provided by the use of internal steam coils. The capacity of the pans may range from a few kilograms up to 100 kg or more, but calculations should be carried out to balance the input and output, the input should not be so large as to allow boiling over and consequent wastage, while the output should not be so small as to allow burning of the

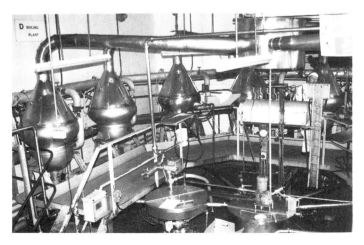

Figure 7.5 Atmospheric boiling pans (courtesy of RHM Ledbury, UK).

jam on to the heated surface of the pan. Frothing of the jam, especially at the beginning of the boil, can present a problem – indeed, it can result in boiling over – but the use of a very small amount of vegetable oil often reduces this. For jelly marmalade, an alternative antifoam such as dimethyl polysiloxane may be used, as the use of vegetable oil can cause cloudiness of the marmalade. The tipping of boiling jam (a somewhat dangerous practice!) from the pan into troughs for subsequent pumping to the filling line is rarely carried out as pans with bottom-opening valves are now largely used.

Preserves may also be boiled under vacuum using either batch (Figure 7.6) or continuous methods as well as by a combination of atmospheric pressure and vacuum. The capacity of batch vacuum cookers also varies from a few kilograms up to several tonnes. Purée type jams may be manufactured on a continuous basis using a plate evaporator while for products with defined fruit pieces, scraped surface heat exchangers operating both at atmospheric pressure and under vacuum are used as evaporators.

After boiling, but prior to filling, the jam is normally held in a reservoir so that quality testing can be carried out to ensure that it is of the right consistency when filled. Total soluble solids, pH and temperature all need to be monitored to ensure compliance with specification, and the consistency of the product of the time of filling will in no small way affect the appearance and texture of the filled and cooled jam.

From the holding reservoir (which is generally gently agitated to ensure an even distribution of the fruit pieces and is often heated to maintain a proper filling temperature), the jam is filled into an appropriate container.

Figure 7.6 Batch vacuum boiling pans (courtesy of RHM Ledbury, UK).

Figure 7.7 Portion pack filling (courtesy of CPC (UK) Ltd).

The size of these will vary from the 20 g portion pack to the 20 tonne bulk tanker for the large bakery user.

Portion packing is an art in itself (Figure 7.7), for the packaging material is normally a plastic that is thermally formed into the receptacle prior to the deposition of the jam, which must not be so hot as to deform the container but sufficiently hot to ensure a good gel formation as well as commercial sterility. These packs are normally heat sealed with foil.

Preserves may be filled into glass jars (capacity from 1 oz up to 2 lb) using volumetric piston fillers (Figure 7.8). These are often of the rotary multihead piston type with filling speeds of 300 or more units per minute. Each head should be considered as a separate filling machine, and machines are available where every head is adjustable. Here, great accuracy can be achieved with an overall average filling weight very close to the target weight (which will have been established by statistical analysis). It is very necessary to ensure that the material being filled is of the proper quality to ensure a good distribution of fruit particles as well as an appropriate gel strength.

Preserves for bakery use are packed in large containers, normally at much lower temperatures to prevent caramelisation of the product. Formulation here is very important so that the gel of the jam in the finished application is of an appropriate texture and strength. It must also be acceptable to the user in the state in which it is delivered, and often these two conditions seem almost irreconcilable. Packs vary from plastic pails (Figure 7.9), sometimes with a heat-sealed foil lid to polythene bag-lined

Figure 7.8 Volumetric jar filler (courtesy of RHM Ledbury, UK).

Figure 7.9 Plastic pail filling (courtesy of Broadheath Foods Ltd).

fibreboard cartons to polylined 200 litre drums to bulk tankers. In the last case, the jam may sometimes be transported in an unfinished condition where the user will, for example, add citric acid in a prescribed amount to achieve a pH on the finished product that will enable the gel to be formed at the required speed.

7.2 Fruits preserved by sugar: glacé fruits

In one sense, this method of preservation of fruit is an extension of the jam-making principle, for in many jams, the producer tries to exhibit large fruit pieces within the gel. With glacé fruits, the object is to maintain the original shape of the fruit while raising the total soluble solids of that fruit to a level that is self-preserving; normally 72% or above (as measured by refractometer). In order to maintain the original fruit shape, it is sometimes necessary to pretreat the fruit to modify its structure. In the case of cherries (the most popular glacé fruit), this is achieved using calcium which, in effect, toughens the fruit by changing the pectin structure.

Cherries provide a good example of the process. After picking, the fruit is stored in a sulphur dioxide solution of sufficient strength to prevent deterioration of the fruit. This solution also contains a calcium salt – some writers have advocated calcium hydroxide while others recommend calcium chloride. After a minimum period of storage and when the texture of the sulphited cherries is considered satisfactory, the processing begins. The first stage is the partial removal of the preserving sulphur dioxide, which is achieved by successive leachings with hot water. This processing has the additional effect of softening the internal structure (this is of benefit not only in the final texture of the fruit but also in facilitating the osmosis by which the sugar enters the fruit).

The cherries are strigged and pitted after they have been size-graded. The pitted cherries are then boiled in water to complete the removal of the sulphur dioxide and the softening operation. If the cherries are to be coloured by the synthetic dye erythrosine (E127), the colouring takes place at this stage. If natural colours are to be used, the colour is added later in the sugaring process. After boiling, the cherries are transferred to a vessel containing a sugar solution at approximately 20%. Over a period of at least seven days, the strength of the sugar solution is increased in stages on a daily basis until a final concentration of between 70 and 74% is achieved. As osmosis takes place, the concentration of the syrup in the vessel decreases. The vessels used for the sugaring are interconnected and syrup of the highest concentration is placed in the final stage vessel. As this decreases in strength, it is passed into the penultimate stage and so on. As the process proceeds, the cherries increase in total solids content and the syrup strength decreases, i.e. the glacé process is a counter-current exchange. The process is outlined in Figure 7.10.

Other soft fruits may be treated similarly except that, in most cases, the initial treatment with sulphur dioxide is not used. Fruits such as apricots, peaches (both of which are normally glacéd 'on stone') and the trimmed halves of pears or slices of pineapple are sometimes used fresh although they may be canned in a heavy syrup for glacéing at a later date.

184   FRUIT PROCESSING

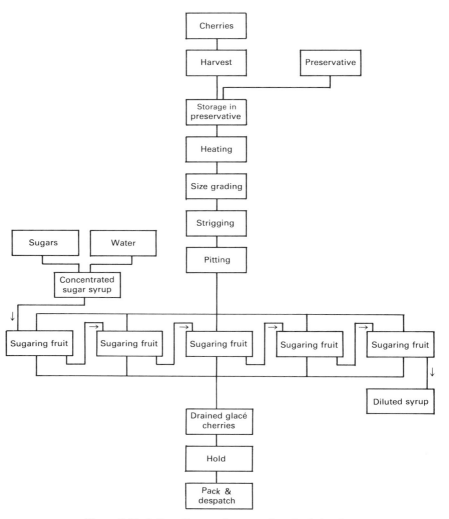

Figure 7.10 A flow diagram for preparing glacé cherries.

A commodity that has been traded for several centuries is sometimes referred to as 'Ginger Sweetmeat' and originated in China. It is, in fact, the rhizome of the ginger plant (*Zingiber officinale*) which has been harvested, trimmed and impregnated with sugar by a similar method to that used for cherries.

The rhizome becomes more fibrous with age and so the time of harvest is important in determining the texture of the finished product. That harvested early in the season is used for sugaring, while the later harvested (and so tougher) rhizomes are dried for flavour extraction or grinding. The harvested rhizomes are stored in vats until required for use, when they are trimmed, cut and sorted into the required shapes. The preservative is removed by leaching and the sugaring process continues by the immersion of the pieces in successively stronger sugar solutions. The glacé (or sugared) ginger is then packed to the customer's specification prior to shipping.

Another commonly available sugared product is citrus peel, normally available as 'cut mixed peel'. Formerly this too was prepared in a similar manner to that used for cherries, using 'half caps' (i.e. a hemisphere of the peel of the fruit). This peel was supplied to the manufacturer packed in brine (a solution of common salt) in large barrels. This peel was de-brined by successive soakings in clean water and tenderised by cooking. The half caps were steeped in successively stronger sugar solutions but here, quite often, the syrup was drained from the peel, strengthened by the addition of more sugar and boiled before being added back to the peel while hot.

Nowadays, the peel is generally treated at source. It is diced using the same machinery as that used for the marmalade shred production, tenderised by pressure cooking and then sugared by cooking with sugar under vacuum. This is a much quicker process. The peel, after sugaring, is drained and packed for subsequent shipment and is generally a mixture of sweet orange and lemon peels. The use of glucose syrup as a partial replacement for sucrose is technically advantageous as it helps in preventing crystallisation and improves the appearance of the product by providing it with a greater sheen than occurs with sucrose alone.

A process sometimes used after the sugar impregnation is that of crystallisation. Here, the analysis of the preserved fruit or peel is important as the amount of inversion of the sugar that has taken place during the processing influences the final texture. Crystallisation may be achieved by coating the preserved fruit or peel with a supersaturated sucrose solution and then applying heat to drive off moisture and so induce the formation of sugar crystals. The use of the supersaturated sugar solution is the traditional method for the making of crystallised citrus peel, but its use, like many of the processes so far discussed, is somewhat of an art, probably because of the 'by hand' nature of the work.

Although not fruit, there are two further items in addition to ginger that merit brief discussion: crystallised flowers and angelica. The second of these is the young stem of the angelica plant (*Angelica archangelica*), an umbelliferous plant introduced from Europe that has become naturalised in parts of the UK. It should not be confused with other inedible umbels! The stems are cut and trimmed and then tenderised and sugared in a similar manner to cherries. They are often coloured green.

Flowers such as violets or rose petals are crystallised in a somewhat different manner. The flowers are harvested and coated in a solution of a gum; often gum arabic is used. The flowers are then dusted in icing sugar and dried: this achieves preservation. Alternatively, the petals are coated with a supersaturated sugar solution and then dried to give rigid crystallised petal-like pieces, which may be used for cake decoration. The syrup used for this process is often coloured to impart the traditional colour to the product. Care must be taken to ensure that the flowers used are edible!

7.3 Fruits preserved by drying

The most well-known fruits preserved by drying are grapes. The various types produce currants, sultanas and raisins. Other fruits are also dried; for example, apricots, pears, bananas and plums (to yield prunes). Whilst the principle is the same for all the fruits, many of the soft fruits are sulphited prior to drying; the processes are elaborated upon below.

7.3.1 Dried vine fruit production

It is an oversimplification to say that currants, raisins, sultanas and the like are dried grapes, although this is essentially the case (Figure 7.11). The history of raisins (the word is here used as a generic term) goes back over 3000 years; it is possible that the discovery was made accidentally where grapes were dried on the vine. Different varieties of grape originated from different regions of the Mediterranean area and each was used to produce a raisin characteristic of that variety. For example, grapes for the Muscat raisin were (and still are) grown in Malaga and Valencia, Spain, while grapes for currants were planted in Corinth and spread to the Zante region of Greece. Persia was the birthplace of the seedless grape from which sultanas are obtained. With time, viticulture spread to the New World as well as to the antipodes, and, today, raisin production is a worldwide occupation.

Practices vary slightly from place to place but one essential requisite is a good climate; adequate sunshine, a suitable ambient temperature and, especially during the drying period, a low rainfall, for the grapes are dried naturally and in the open air.

There are slight differences in terminology from the various growing areas; currants are universally derived from the small dark seedless Zante type grape (the so-called Black Corinth of California) although other grapes with similar characteristics (e.g. the Australian Carina) are used. In Australia, sultanas are obtained from the Thompson Seedless grape as well as the Sultana. In California, raisins are also obtained from the Thompson Seedless while the Sultana grape yields sultanas as it does in Europe. Muscat raisins are known as such in Europe and America, while in

PRESERVES, FLAVOURINGS AND DRIED FRUIT

Figure 7.11 Raisins (courtesy of the California Raisin Advisory Board).

Australia they are simply known as Raisins. This last category is made from large grapes containing seeds, which may or may not be removed mechanically during the manufacturing process after drying. Muscat raisins have a much sweeter taste than the other types.

The processes after grape harvesting vary slightly from place to place, sometimes the grapes are dried naturally without any further treatment or they are pretreated to speed up the drying process. The 'natural' raisins of California are produced without intermediate treatment, being spread on paper sheets laid between the rows of vines to dry (Figure 7.12). In two to three weeks, this reduces the moisture content of the berries from over 70% to about 15% at which stage the paper sheets are rolled with the dried grapes inside to produce 'bundles'. These are left for several more days for further drying and to equilibrate the moisture contents of the grapes within the bundle. The grapes are then stored in large containers (which further assists with this equilibration) at the packing plant. Prior to removal from

Figure 7.12 Drying raisins (courtesy of the California Raisin Advisory Board).

the bulk bins, samples are taken for quality approval (United States Food & Drugs Authority Standards are applied here). The raisins now pass through a cleaning procedure to remove leaves and large pieces of stems, with the small capstems being removed by a special machine in which the fruit is cascaded through a spinning grooved rubber-coated cone. Further cleaning is achieved by vacuum separator treatments, where the lighter stem material is removed. Following washing and grading, the raisins are packed into the packs from a few grams for a snack pack to bulk packs of around 12.5 kg for use by other food manufacturers. A basic flow diagram for the Californian process is given in Figure 7.13.

In Turkey, woven polypropylene sheets are used, and these are laid on cement or clay beds, often slightly higher than the surrounding ground to reduce the possibility of contamination of the fruit by stones etc. In Australia, drying racks are constructed. These consist of 8 to 12 tiers made of wire netting within the rack which itself is about 2.5 metres high. The grapes are spread on the mesh at about one bunch deep for drying, either naturally or by the accelerated method outlined below (here the individual layers of grapes on the rack are sprayed with the alkaline oil treatment). There is some mechanical drying of the fruit carried out, often by placing racks of grapes in a tunnel and blowing heated air through. When dried to the correct moisture content, the raisins pass through the cleaning and packing processes as outlined above (Figure 7.14).

For some products, extra process(es) may be introduced. In order to speed up the drying process, the grapes may be sprayed after harvesting

PRESERVES, FLAVOURINGS AND DRIED FRUIT 189

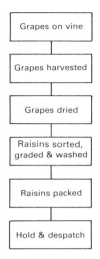

Figure 7.13 A flow diagram for the Californian process to produce raisins.

Figure 7.14 Bulk packing of raisins (courtesy of the California Raisin Advisory Board).

with, or dipped in, a solution of potassium carbonate (between 2.5% (w/w) as used in Australia and 4.5% (w/w) as used in Turkey) containing 'dipping oil', the composition of which varies from place to place. In Australia, it is a mixture of the ethyl esters of fatty acids and free oleic acid used at 1.5% (w/w), while one establishment in Turkey uses 0.5% olive oil. This can reduce the drying period for sultanas from four to five weeks to 8–14 days. It was originally believed that this process removed the waxy bloom from the surface of the grape, but this is not the case. It is now thought that it modifies the structure of this waxy coat to allow greater moisture permeability and hence allows dehydration to proceed more quickly. It also appears to make the grape more transparent to infrared rays, allowing a better radiant heat uptake. The effect of the treatment is reversed by washing.

Because of their larger size, the Muscat grapes are normally alkaline treated before drying, and the treatment tends to be a little more severe than with the smaller types of grape. A relatively small amount of Muscat grape is mechanically dried without the alkali pretreatment and is handled delicately to minimise damage and to preserve the surface bloom. The resulting raisins are called 'Muscatels' and are used for high-class outlets, such as health food stores.

Another process that may be applied to some sultanas (and raisins in California) is their treatment with sulphur dioxide. This bleaches the fruit to give a much more golden produce but can leave a residual amount of the preservative in the fruit (up to 2000 mg/kg).

7.3.2 Dried tree fruit production

The drying methods for tree fruits are essentially the same as those for vine fruits, i.e. the sun-drying of prepared material. Soft fruits such as apricots, pears, bananas and plums are dried, as are dates and figs. Some of these fruits have to be treated with sulphur dioxide before drying to prevent spoilage, and this particularly applies to sun-dried cut fruits, such as apricot halves.

Fruit for drying is harvested in the normal way with care to minimise damage to the fruits. Dates and figs are dried in a similar manner to grapes, that is on prepared cement or clay beds, although the use of trays is quite possible. Stone fruits such as apricots may be dried whole on stone, whole after having the stone removed or as stoned halves. Plums (to produce prunes) are normally dried whole on stone. Fruit for drying should be mature and fully ripe: underripe fruit does not produce such a flavourful commodity.

In Australia, fruit is dried, as with grapes, on trays but these trays are of a softwood construction (hardwood imparts a staining to the fruit) and are stackable. The fruit is placed in a single layer on the trays for drying. At this stage some of the fruit types are treated with sulphur dioxide. Apart from preserving the fruit, this process softens the fruit tissues, facilitating a faster loss of moisture during drying. It also has an antioxidant effect and thus prevents enzymic browning of the fruit.

For those fruits that are to be treated with sulphur dioxide, the process is quite simple. Trays of fruit are formed into stacks that are placed in purpose-built fumigation chambers which are only a little larger than the stacks themselves. Sulphur is then burned in a draught channel under the chamber and the gas enters the chamber and treats the fruit. For legislative reasons (if for no other!) control on the degree of treatment is necessary. After fumigation, the fruit is sun-dried to a moisture content that is adequate for self-preservation when combined with the sulphur dioxide.

Some fruits, such as plums (for prunes), are dried without treatment with sulphur dioxide, and, here, the drying is carried out artificially using a process similar to that mentioned above for some types of grape.

The drying of fruit is a complex subject and a whole book (rather than a small part of a single chapter) would be needed to do it complete justice.

7.4 Flavourings from fruits

Many flavourings in present use are called 'natural' or 'nature-identical' (similar but synonymous terms are also used). Many of these comprise concentrated fruit juice (which in many circumstances is an excellent flavouring but which, in a list of ingredients, need not be referred to as such) or concentrated fruit juices that have been 'boosted'.

The subject of fruit juice concentration is dealt with in chapter 4. During the evaporation process vapours are generated that contain many of the essential flavouring compounds. These vapours are collected and concentrated by fractional distillation before being added back into the concentrate. The recycling may be at the natural ratio (i.e. the amount recycled is the same amount as that collected); sometimes the volatiles are added back at higher than natural ratios to produce a fortified concentrate.

Sophisticated chemical analysis is now able to identify these flavouring compounds, many of which have been synthesised for use in the manufacture of so-called 'nature-identical' materials.

One of the most important sources of flavouring is the essential oil of the citrus fruits. These are extracted from the peels of the fruits (see chapter 4) and are widely used, not only in food flavours.

7.5 Tomato purée

Is it a fruit or is it a vegetable? The humble tomato seems to fall between two stools, for while botanically it is undoubtedly a fruit, it is specifically excluded from some fruit juice legislation (for example the UK Fruit Juices and Fruit Nectars Regulations of 1977). It is often used more as a vegetable than a fruit. Tomatoes can be conveniently divided into two categories: the round plum type and the long Italian type. This latter type is more susceptible to spoilage by *Bacillus thermoacidurans*, and, for its treatment, higher processing temperatures are required. The term tomato juice is taken here as being the material obtained by sieving tomatoes after they have been chopped and preliminarily treated (see below). Tomato purée is taken as being concentrated tomato juice.

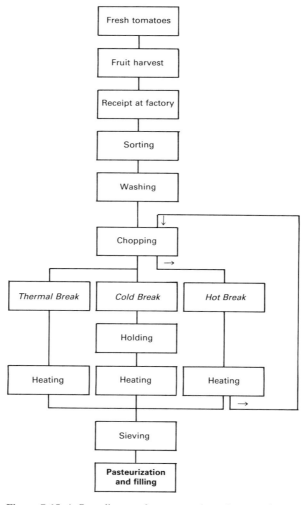

Figure 7.15 A flow diagram for preparation of tomato juice.

Much of the plant and machinery developed for tomato processing is very sophisticated, and only an outline is given here. Before the fruit is received at the factory, it is checked for quality, particularly for mould growth, an important factor affecting the quality of the end-product.

At the factory, the fruit is checked again and sorted to remove defective fruit and other extraneous material. It is then washed with potable water before being crushed or chopped (this is often termed 'breaking' and is linked to one of three types of preliminary treatment). The treatment used depends on the intended use for the end-product.

The 'cold break' process involves holding the crushed or chopped fruit at ambient temperature for a period in order to destroy the natural pectin by enzyme action. This lowers the viscosity of the tomato juice and allows for higher concentrations of tomato purée to be achieved. In the 'thermal break', the fruit is chopped cold and the resulting pulp is heated immediately to 90°C before the juice is extracted. The time between chopping and heating governs the amount of pectin breakdown and so governs the viscosity of the end-product. Hot-break juice is achieved by heating the fruit to 90°C before chopping and pulping, the material being recirculated in order to ensure that the enzymes are inactivated before the pectin is broken down. The heating also releases a gummy substance that surrounds the seed to give a high viscosity end-product, suitable for the ketchup industry.

The heated tomato pulp is then sieved to remove the seeds and skin as well as other fibrous parts of the fruit and any stalk that may be present. The sieving is achieved by passing the pulp through a series of sieves each with smaller apertures than the last, the final one having an opening of 0.4 mm. Figure 7.15 shows a flow diagram from fresh fruit to the juice.

Following the sieving operations, the 'juice' is transferred to the evaporation equipment for concentration. The type of evaporator used will depend on the product required to some extent but is almost always carried out under vacuum. The scraped surface type is mainly used, often with double or triple effect. The use of scraped surface plant allows a concentrate of a much higher viscosity to be prepared. It is not usual to recover volatile aromas during tomato processing. Much concentrate is now homogenised, while a customer specification may require the presence of added salt. Reference to the plant manufacturers' data sheets will provide plant throughputs.

After concentration, the purée may be canned, when it will need to be heated to a suitably high temperature to achieve sterility, followed by rapid cooling to prevent caramelisation. Alternatively, the purée may be handled aseptically and bulk filled into containers of 200 kg or more (Figure 7.16). Here the plant used for the evaporation is maintained in a sterile condition and the purée is cooled while still maintaining that sterility

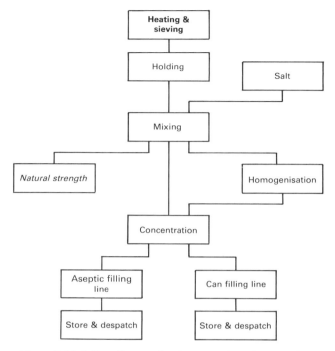

Figure 7.16 A flow diagram for the processing of tomato juice.

for filling into sterile containers. To improve its shelflife, such purée is often chilled at the completion of the processing.

The heated sieved juice may be pasteurised as single-strength material for filling into a variety of packs, the evaporation temperatures achieved during concentration must be sufficient to pasteurise the purée.

Acknowledgements

The author gratefully acknowledges assistance given to him by the following Companies and other bodies, together with their staff.

APV Baker Ltd, Crawley, West Sussex RH10 2QB, UK.
Broadheath Foods Ltd, Lower Broadheath, Worcester WR2 6RF, UK.
Buderim Ginger (UK) Ltd, Croydon CR0 4NH, UK and Buderim Australian Ginger, Yandina, Queensland, Australia.
CSIRO, N. Ryde, N.S.W. 2113, Australia.
California Raisin Advisory Board Food Technology Program, PO Box 281525, San Francisco 94128–1525, California, USA.
CPC UK Ltd, Esher KT10 9PN, UK.

PRESERVES, FLAVOURINGS AND DRIED FRUIT 195

Pagmat, S.A. Atatürk Cad., No. 150/2 (35210) Izmir, Turkey.
RHM (Ledbury), Ledbury HR8 2JT, UK.
Thomas J. Payne Market Development, 3242 Jones Court, N.W. Georgetown, Washington DC, USA.
Tesco Stores Ltd, Cheshunt EN8 9SL, UK.
Campden Food and Drink Research Association, especially Members of the Library Staff, Chipping Campden GL55 6LD, UK.

Further reading

Atkinson, F.C. & Strachan, C.C. (1941). *Candying of Fruit in British Columbia with special reference to Cherries.* Contribution No. 573, Horticultural Division. Experimental Farms Service, Experimental Station, Summerland, BC, Canada Department of Agriculture.

Atkinson, F.C. & Strachan, C.C. *Preparation of Candied Fruits and Related Products*, Part 1. Fruit and Vegetable Processing Laboratory, Experimental Station, Summerland, BC, Canada Department of Agriculture.

Campbell, C.H. (1937). *Campbell's Book.* Vance Publishing, New York.

Goose, P.R. & Binstead, R. (1964). *Tomato Paste and other Tomato Products.* Food Trade Press, London.

McBean, D.M. (1976). *Drying and Processing Tree Fruits (Division of Food Research Circular No. 10).* Commonwealth Scientific and Industrial Research Organisation, NSW, Australia.

Ranken, M.D. (ed.) (1988). *Food Industries Manual*, 22nd edn. Blackie, Glasgow.

Ranken, M.D. & Kill (ed.) (1993). *Food Industries Manual*, 23rd edn. Blackie, Glasgow.

Rausch, G.H. (1950 and 1965), *Jam Manufacture.* Leonard Hill (Blackie, Glasgow).

Various (1982). *Grape Drying in Australia.* Dried Fruits Processing Committee, Commonwealth Scientific and Industrial Research Organisation, N. Ryde, NSW Australia.

Various UK Legislation Statutory Instruments, Her Majesty's Stationery Office, London.

8 The by-products of fruit processing
R. COHN and A.L. COHN

8.1 Introduction

The production of by-products from the waste of the citrus processing industry has increased in recent decades for the following principal reasons:

1. The increase in total quantity of fruit used for juice manufacture.
2. The economic importance of the by-products to the juice producer because they increase profitability.
3. The need to overcome the environmental problems caused by waste material accumulating in each plant.

The statistics of the Brazilian and USA citrus industry in 1991/2 serve as an excellent example. The total fruit supply to the citrus industry in 1991/2 in Brazil was 8 107 000 tonnes of which about 4 000 000 tonnes was waste material, whilst in the USA, 7 155 000 tonnes of raw fruit was used and gave rise to about 3 300 000 tonnes waste material. There are large differences between the different fruit varieties in the amount of waste per tonne from processed fruit. In the apple industry, for example, waste is 10–15%, which compares with about 50% in the citrus industry and up to 70% with some tropical fruits. Nearly all the waste from the citrus industry can be successfully converted into animal food or fodder products.

8.2 By-products of the citrus industry

Citrus fruit production can be divided depending where it arises in the juice extraction process:

1. juice and juice cells, which form about 40–55% of the fruit
2. peel (flavedo) and rag (albedo), which constitute about 45–60% of the fruit; the flavedo contains the essential oils and the carotenoid pigments; the albedo contains cellulose, pectins and flavonoids.

Table 8.1 shows the different products which are produced from citrus fruit. The main by-products from the endocarp, or inner part of the fruit, are juice cells and pulpwash concentrate.

Table 8.1 Products of the citrus industry

Juice and cells	Peel and rag	Essential oil
Concentrated juice	Pectin	Cold-pressed oil
Juice	Cloudy concentrate	Terpenes
Premium pulp	Hesperidin	Concentrated oil
Pulpwash concentrate	Naringin	Distilled oil
Dehydrated cells	Dried peel	
Water and oil phase volatiles	Molasses	
	Alcohol	
	Natural colour	

8.2.1 Citrus premium pulp (juice cells)

Juice cells are separated from the juice by sieving with variable pressure through a rotating drum finisher. Embryonic pips and small pieces of peel are separated by a cyclone installed in the juice production line. Typically, cells are pasteurised and cooled immediately after sieving in order to destroy pectic enzyme activity and to prevent fermentation. The cells contain natural juice and have the same Brix value as the juice. This product is used mainly to improve flavour and mouthfeel in single-strength juice made from reconstituted concentrate. Cells may be frozen or aseptically packed into drums or special cartons.

Washed pulp and pulpwash concentrate. An alternative industrial possibility of utilising the juice retained in citrus pulp is by extraction with water. This is usually done by a multistage counter-current extraction system, using juice extractors for pressing the washed cells. Depending on the number of stages, it is possible to produce a low Brix (3–6%) pulpwash liquid as well as washed pulp. The washed pulp is pasteurised, cooled and packed in the same way as premium pulp. It has a low Brix, a high percentage of broken cells and a weak flavour. This product is mainly used in drinks to increase 'body' and improve appearance.

Because of its high pectin content, the liquid recovered from washed cells is very viscous and difficult to concentrate. Therefore, most producers treat the extract with pectolytic enzymes at 45°C batchwise or by a continuous process. After pasteurisation and centrifugation (for pulp reduction), the pulpwash extract is concentrated under vacuum to 58–64°B. Orange and grapefruit pulpwash concentrate is packed into drums and frozen. Composition of orange pulpwash concentrate from different origins is shown in Table 8.2.

The quality of the product is characteristic of the fruit as well as the technology used in its production. Low pulp content, good cloud stability and a neutral (i.e. absence of bitter) taste are requisites for high quality. Valencia orange will give a product superior in colour and taste to an early

Table 8.2 Composition of commercial pulpwash concentrate

	Orange		Grapefruit
	USA	Brazil	USA
TSS (B)	55.0	63.5	60.0
Citric acid (anhydrous) (% w/w)	3.2	4.5	4.3
Formol No. (100 ml)	23	21.8	31.2
Total sugar after inversion (% w/w)	39.8	45.3	41.4
Potassium (mg/kg)	1953	1700	1750
Pulp in juice (%) 1500 r.p.m./10 min	2.0	8.0	2.0

Table 8.3 Composition of dehydrated washed cells (%)

Moisture	Fat	Protein	Ash	Crude fibre	Pectin	Carbohydrate	Total
11.7	2.1	7.5	2.5	9.8	29.0	37.4	100

Ferguson and Fox (1978).

season Navel. As the pulpwash concentrate is prepared by washing the cells with water, the legislation in some countries does not permit their subsequent use in natural juices, only in drinks.

Dehydrated citrus cells. The market for frozen citrus cells is limited and there are large quantities of this raw material, about 10–15% of the extracted juice, depending on the equipment used. Therefore, the industry is looking for additional products that can be made from cells. Additionally, if the surplus cells are dried together with peel residues, part of them is lost as dust leading to air pollution.

By drum drying the washed pulp, an edible dietary fibre product is produced that can be used in several ways. The composition of these fibres is shown in Table 8.3. The dried cells have a high pectin and crude fibre content which makes them a valuable dietary product. Research carried out recently tends to support claims for their gastrointestinal function in preventing diseases of the colon, lowering cholesterol and preventing heart diseases (Baker, 1993).

Dried juice cells have high water and fat absorption ability. One part of dried cells can absorb 10–15 parts of water and 3–5 parts of fat. This opens opportunities for the baking industry for dried juice mixes and meat products (Kesterson and Braddock, 1973).

8.2.2 Products prepared from peel and rag

The peel and rag of citrus fruit are used mainly as cattle feed, but as a result of considerable investment in industrial research, important edible products have been developed from them. Essential oils extracted from the peel are one of the sources for aroma development and terpenes have valuable uses in the chemical industry. The composition of the peel (Table 8.4) shows the importance of sugar and pectin. The pectin industry uses increasing amounts of citrus peel as raw material instead of the traditional apple waste. New developments in this field especially the process of debittering have enabled the utilisation of a large part of the sugars and other soluble ingredients, which can be used in the drink industry. The bioflavonoids, hesperidin and naringin, have found a place in the pharmaceutical industry. The following paragraphs describe the industrial production of these products.

Production of 'Special Concentrate' from peel extract. Shredded peel and rag are used as raw material. Their extraction with water is carried out in a number of stages, either batchwise or with a continuous counter-current diffusion process that includes heating. To reduce viscosity, most manufacturers add specific pectolytic enzymes during or after extraction and adjust the temperature, time and pH to achieve optimal activity. The mixture of pectolytic enzymes may include polygalacturonase, lyase, transeliminase as well as cellulase (Pilnik and Voregen, 1991). An alternative process is to heat the peel–water mixture to 98°C prior to enzyme treatment; this eliminates the undesirable action of the pectin esterase present in the citrus peel, which results in loss of cloud stability. At the end of the extraction process, the liquid is separated from the peel by a decanter. The extract varies from 3 to 6°B depending on the number of extraction stages. Extraction is followed by pasteurisation to inactivate any enzyme activity and centrifugation to reduce pulp content as much as possible.

The peel liquid is finally concentrated *in vacuo* to 50 or 60°B. The colour

Table 8.4 Composition of Israeli orange and grapefruit peel

	Orange		Grapefruit	
	Mean	Range	Mean	Range
TSS	12.5	9.4–17.7	10.9	8.4–13.4
Total sugar (% w/w)	8.0	5.5–10.5	6.7	4.7–8.4
Ash (% w/w)	0.54	0.44–0.9	0.57	0.49–0.73
Potassium (mg/kg)	940	600–1440	1077	815–1340
Total pectin (g/kg)	3.65	1.7–7.3	2.95	0.27–5.63
Hesperidin (g/kg)	18.5	14.6–23.9		
Naringin (g/kg)			12.3	7.3–17.3
Formol No. (ml NaOH, 0.1 N/100 g)	27.5	18.3–36.7	22.3	12.0–33.6

Cohen *et al.* (1984).

of the product depends on the carotenoid content of the peel used for extraction. Early fruit gives light coloured concentrate (30 p.p.m. total carotenoids), Valencia about 50 p.p.m. and easy peel varieties up to 80 p.p.m. carotenoid. These special concentrates are all very bitter and can be added only in small quantities to a drink. They contribute a stable cloud, natural colour and contain the bioflavonoids of the peel. The turbidity (cloud intensity) of the orange peel extract is a result of the interaction of the pectin, hesperidin and protein (Ben-Shalom and Pinto, 1986; Kanner, Ben-Shalom and Shomer, 1982). No interaction occurs between naringin and pectin and, therefore, grapefruit peel extract is not very turbid. The high naringin content of grapefruit peel extract often results in precipitation of naringin crystals in the concentrate. A typical flow diagram is shown in Figure 8.1. A similar process, but using different raw material, is described by Ragonese (1991).

Debittering of citrus by-products. All the citrus peel extracts are extremely bitter because of the very high limonin content. For example, orange peel contains 300–400 p.p.m. limonin, and the capillary membranes about 300 p.p.m. on a wet basis. The threshold of bitter taste resulting from limonin in orange juice is about 6 p.p.m. Some consumers even find 1 p.p.m. quite bitter. Considerable research has been carried out to find ways of reducing the bitterness of orange and grapefruit products and many commercial chemical companies now offer complete debittering plants. In the USA, the main effort has been concentrated on debittering Navel orange juice, in Italy and Israel on debittering cloudy peel concentrate extracted from peels. Grapefruit products can only be debittered by reducing both limonin and naringin. The structural formulas of limonin and naringin are shown in Figure 8.2.

The first material used for this debittering by the food industry was activated carbon. However, this method is no longer used because many vital juice components (such as ascorbic acid or the carotenoids) are also adsorbed by carbon.

The second method used specific enzymes. Naringinase free from pectinase can be used for reducing the bitter taste of grapefruit products. The optimum conditions are a temperature of 45–50°C and the time varies depending on the amount of naringin present. The enzyme limonoate dehydrogenase has been evaluated for decomposing limonin (Hasegawa, Brewster and Maier, 1973).

The most effective debittering technique has been introduced only recently to the citrus industry and involves the adsorption of limonin and naringin onto special resins using ion-exchange equipment and techniques. The US Food and Drug Administration has permitted this method for debittering natural Naval orange juice.

The first step is to reduce the pulp content of the peel extract by

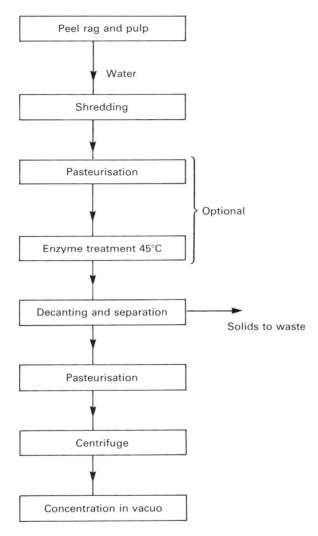

Figure 8.1 Production of cloudy concentrate from peel extract.

centrifugation to less than 1%; this prevents clogging of the resin. Resins adsorb limonin and naringin without affecting the other juice constituents.

It is possible to use ultrafiltration for separating the fluid from the pulp, but this is an additional investment. However, without ultrafiltration the column must be backwashed from time to time to remove residual pulp from the resin.

Undebittered pulp can be added to debittered juice at the end of the process in order to recover the original cloud intensity of the peel extract.

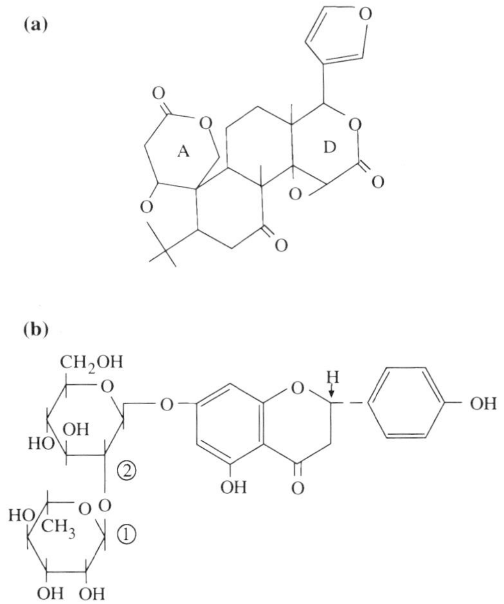

Figure 8.2 Structure of (a) limonin and (b) naringin.

Most of the industrial equipment consists of alternating twin columns, allowing juice treatment and regeneration to proceed simultaneously in the different columns. The system can be fully automated, with regeneration using diluted caustic soda.

In big installations, up to 10 000 litres of extract per hour can be debittered. The final step of the operation is vacuum concentration. Two different concentrates are produced; one clear the other cloudy and both have their uses in the drinks industry.

In Italy, a special concentrate is produced from whole lemon or oranges of inferior juice quality. Torchiato press residue is diluted with water, heated and enzyme treated. After pasteurisation, the juice is centrifuged and debittered and forms a valued product for drink bases.

Hesperidin and naringin production. The production of citrus flavonoids began when it was realised that they possessed biological activity. Research in recent years has moved from a consideration of the effect of bioflavonoids on blood capillaries to a more central role relating to their general function in human health (Robbins, 1980).

Hesperidin is the bioflavonoid mainly present in oranges and mandarins, *Citrus sinensis* and *Citrus reticulata*, as well as in all lemon varieties (Figure 8.3). In Shamuti orange juice, the concentration of hesperidin is between

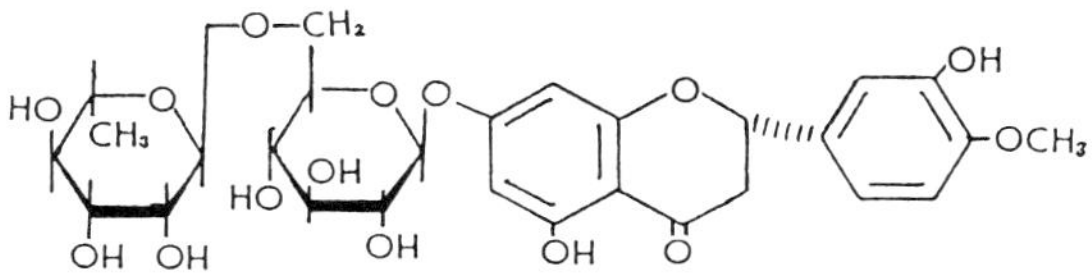

Figure 8.3 Structure of hesperidin.

0.8–2.0 g/kg and in the peel it is 13–24 g/kg. Hesperidin is practically insoluble in water at acid pH but is soluble in alkaline media. To extract hesperidin from orange peel, the shredded peel is mixed thoroughly with a dilute lime solution to a final pH of 8.0 by mixing in a tank or screw conveyor for up to 30 minutes. The peels are pressed in a screw press, as used in the production of dried citrus peel. The alkaline liquid is screened and then filtered with the assistance of filter aid in a filter press. The resulting clear liquid is acidified in a tank to a pH below 3.0 and hesperidin crystals are separated from the liquid by centrifugation. The remaining peel extract of about 11°B and the pressed peels can be used in the production of molasses and dried cattle feed. The main market for hesperidin is in the pharmaceutical industry. The concentration of naringin in grapefruit peel is between 0.7 and 17.0 g/kg. Naringin is produced in a similar way to hesperidin, but the crystallisation process is more difficult because of its greater solubility in water, even in acid solution. Therefore, concentration of the liquid and cold storage are recommended to enhance the process. Naringin is used in drinks to enhance the bitter taste and it has been examined also as a raw material for further processing to an artificial sweetener.

8.2.3 Citrus oils

Essential oils are extracted from all varieties of citrus as prices are currently attractive and the production does not demand great investment. World production, according to Wright (1990), is about 16 000 tonnes orange oil, 2 500 tonnes lemon oil and 180 tonnes grapefruit oil. Lime and bergamot are among the highest priced oils. The main market for essential oils is for the flavouring and drink industry.

Cold pressed oils. Citrus oil develops in receptacles on the flavedo of the peel, which is covered by a protective wax. The quality and quantity depend on the variety and ripeness of the fruit, as well as the location of the grove. For example, oil from Valencia oranges is considered to be the best quality amongst orange oils and Italian lemon and green mandarin oils are regarded as being superior to oils from other sources. The method of extraction is also of vital importance. The yield of essential oil from the fresh fruit is about 0.4% in lemon, between 0.2 and 0.4% in orange and 0.2% in grapefruit. The main steps in the production of cold pressed essential citrus oil are:

1. Extraction of the oil by mechanical pressure on the oil vesicles
2. Separating the oil from the peel by water spray
3. Centrifugation of the water/oil emulsion
4. Clarification of the crude oil in a centrifuge
5. Settling of the waxy crystals present in the oil by holding at low temperatures for several weeks (winterising)
6. Separating the wax from the oil by centrifugation before packing into drums.

The extraction method used by each producer depends primarily on the type of juice extractor used in the plant. The FMC extractor, which simultaneously separates oil from the fruit during the juice extraction process, is used in most citrus-producing countries. The Brown juice extractor has two different oil extraction methods: (i) the Brown oil extractor is used before juice extraction, and (ii) the Brown peel shaver is used after juice extraction. All three American systems produce good-quality oil. The different systems are described in detail by Redd and Hendrix (1995).

In Italy, different oil extraction systems have been developed and are adapted to the unique fruit varieties grown there. Two of them are still in use together with the American machines (Ragonese, 1991). The Pelatrice system is used by the Indelicato and Speciale oil extractor. The whole fruit is rasped by special rotating rollers which carry the fruit along a closed box where water sprays wash the oil released by the rasping process. The Torchi system replaces the older Sfumari system. This oil extractor is used for peel as well as small whole fruit. The oil is extracted by two contra-rotating helices with increasing presure along the whole length of the oil press. Also half lemon peels can be treated in this way after they have first passed through a Birillatrice juice extractor. Usually the process is in two stages; the first using low pressure, the second high pressure. The water–oil emulsion from the second stage is difficult to separate because of its high pectin content. The oil produced by this system is of good quality and the yield is high.

Clarification of the oil and waste water treatment. After extraction, the water–oil emulsion passes to a desludging centrifuge producing crude oil and water still containing small amounts of oil, pectin and sugars. In some extraction systems, such as the FMC, the water phase is recirculated into the water spray system to increase the oil yield and reduce the quantities of water that go to waste. Citrus oils act as bacteriocides and will interfere in the waste water treatment; they should be removed as far as possible by recirculation. The Brix of the water rises and can be added to the peel press water and used for the production of molasses. In this way the total solids of the waste water as well as the oil are removed in an environmentally sensitive way.

The crude oil is clarified in a desludger and transferred to settling tanks or drums for dewaxing. The lower the temperature, the faster the settling of the wax, which should take place below 0°C. After several weeks the oil is passed through a polishing centrifuge for dewaxing. Essential oils are inflammable and must be handled carefully. The oil is readily oxidised and, therefore, should be kept in full containers at low temperatures. Although cold-pressed oil contains natural antioxidants, some producers add butylated hydroxytoluene (BHT) or butylated hydroxyanisole (BHA) to prolong the shelf life of the oil.

Aroma and oil from juice. The excellent flavour of freshly pressed juice is caused by small quantities of oil (0.02–0.03%) and water-soluble aromatics (water phase) present in the juice. During the process of vacuum concentration, these are lost by evaporation. In modern juice plants, various aroma recovery systems have been installed, which strip the oil in the first step of concentration, usually after pasteurisation. The oil and water phase are kept separately at low temperatures. They can be added back at a later date to the concentrate to enhance the flavour of the reconstituted juice. Recovered aromas are also used by flavour houses to produce soluble citrus aromas of excellent quality. A detailed explanation of aroma recovery equipment is given by Redd and Hendrix (1993).

Distilled oils and their fractionation. Cold-pressed essential oils can be distilled to produce different aroma fractions for use in the drinks industry. The largest fraction is d-limonene, which comprises up to 95% orange oil and 90% lemon oil, but its contribution to flavour is limited. The important fractions are water-soluble esters and aldehydes, which contribute the fruity note to drinks. A great part of lime and lemon flavour is produced by distillation of cold-pressed oils. A detailed discussion of this subject is given by Redd (1993).

Concentrated oil. Concentrated oils are produced by specialised flavour houses from different cold-pressed citrus oils. The terpenes are partly removed in the process, depending on the degree of concentration (1:5 to 1:20). Concentrated oils are less sensitive to oxidation and have a pleasant flavour; this is the result of the removal of a large proportion of the terpenes, which are insoluble in water. Therefore the formation of oil rings in the drinks is minimised. Concentrated mandarin oil (*Citrus reticulata*) can contribute to colour as well as flavour, as it contains dark coloured carotenoids, such as cryptoxanthin, which are not present in *Citrus sinensis*. When concentrated up to 4000 p.p.m. in the oil these substances can be used as a natural colourant for drinks.

Terpenes. The press liquor produced from peel in the drying process contains a considerable amount of peel oil. In the first stage of molasses production, this oil is stripped and collected in tanks. The terpenes produced in this way have many uses in the chemical and cosmetic industry. Terpenes are excellent general purpose organic solvents.

8.2.4 Comminuted juices

Comminuted juices were introduced by the British Soft Drinks Industry soon after the Second World War. The main supplier of orange and grapefruit communited juices is Israel. Communited lemon juices are produced in Italy because of the superior quality of the lemons grown there. Originally the whole citrus fruit was crushed, passed through a finisher, milled, pasteurised, cooled and stored, preserved. Only 3–4% of this mixture in a final drink is needed to contribute a fresh citrus flavour as well as a cloudy appearance to the drink. This cloud is caused by the presence of essential oil and peel substances in the comminuted juice.

To simplify standardisation of the product, the industry now uses a different method of production. By blending different fruit components according to agreed formulae, many products can be produced to meet client requests. Oil and peel content can be changed and by adding concentrated juice or cloudy peel the concentration of the comminuted juice can be raised to 50°B. In this way, less packing material is used and shipping costs are reduced.

Most comminuted juices are packed in large pastic containers and usually preserved with sulphur dioxide, benzoic acid or both. For aseptically packed comminuted juices, special plastic sacks with an aluminium foil barrier are used. Figure 8.4 gives the flow sheet of the production process.

Most whole citrus drinks on the British market contain up to 15% single-strength comminuted juice and are diluted to taste with water.

8.2.5 Dried citrus peel

Many of the citrus juice processing plants have such a high throughput that disposal of the wet peel directly to farmers is not possible. In the USA and Brazil, nearly all of the waste from the plant is dried and sold as animal feed pellets. Several manufacturers provide the industry with equipment and layouts for the production of dried peel for animal feed as well as for pectin. The general outlay of such a process is as follows:

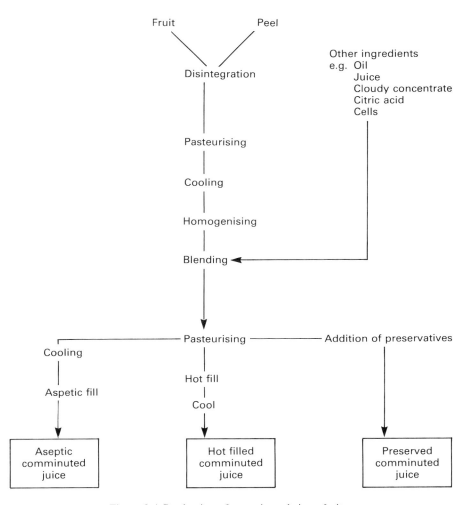

Figure 8.4 Production of comminuted citrus fruit.

1. The waste material is collected in silos and held there for not more than 4 hours. The moisture content at this stage is between 82 and 84%.
2. The peel is shredded.
3. Addition of 0.1% to 0.25% lime (calcium hydroxide) produces a pH of 5.5 to 7.0 and this is held for 10 to 20 minutes. The higher the pH, the shorter the retention time required, but more scaling in the evaporator of the press liquor should be expected. Lower pH is preferred especially if the press liquor is to be used for fermentation to produce alcohol.
4. Pressing gives about 65% press cake with approximately 75% moisture

content and 35% press liquor. The press liquor is concentrated in a multieffect evaporator to produce molasses, using the waste heat of the exhaust gases from the dryer. The molasses are either used as raw material for the production of fuel-grade alcohol or mixed with the press cake.
5. Either the press cake or its mixture with the molasses is fed into a rotating drum drier, reducing the moisture content to 10–15%.
6. Finally, dried peel is passed through a cooling tunnel before being bagged or transported to a storage bin. Most peels are pelletized in order to reduce the volume during storage and transportation as well as to facilitate the feeding of animals. All dried peels should preferably have a moisture content not higher than 10% and be stored in a cool place that, if necessary, is ventilated with chilled air.

The nutrients of the dried citrus peel fluctuate according to the type and degree of ripeness of the fruit, as well as environmental conditions. Taking average figures of several available data, the values in percentage on a moisture-free basis are typically: crude protein, 6.8%; fat, 4.7%; crude fibre, 13.4%; ash, 6.6%; nitrogen-free extract, 68.5%. Comparative values for crude protein and crude fibre, respectively, in other meals are: 49.6% and 7% for solvent-extracted soya bean seed, and 44.8% and 13% for cotton seed (Anon, 1987).

In comparison with other types of feeding material, dried citrus waste is of inferior quality as far as protein content is concerned and is consequently lower in price.

The production of fuel-grade alcohol, which during recent years has played an important role, especially in the Brazilian industry, assists in counterbalancing the economic disadvantages of disposal of the waste from the citrus industry.

8.3 Natural colour extraction from fruit waste

From ancient times, natural colours have been used to give food a more attractive appearance. The addition of artificial colours to food products has become a controversial subject in the last few years and now many such colours have been restricted in use. Even natural and nature-identical colours (like beta-carotene) must now be declared on product labels in most countries.

The extraction of colour from fruit waste, therefore, has been intensively studied and ways of industrial application suggested. Two colours are of the utmost importance for the juice and drink industry; the orange-red colour of the carotenoids and the dark red-blue colour of the anthocyanins.

8.3.1 Extraction of colour from citrus peels

About 70% of the carotenoids of the whole fruit are concentrated in the flavedo. The main components are xanthophylls, but there are many other carotenoids in citrus varieties and their dependence on ripeness, soil condition and other factors has been discussed by Gross (1997). The extraction of the colour from citrus peel is not easy as the carotenoids are not water soluble. Therefore, only a few methods of extraction have reached industrial application.

Several research institutes have suggested different ways of colour extraction from wet peel (Ting and Hendrickson, 1969; Wilson, Bissett and Berry, 1971). A research team at the Volcani center in Israel (Kanner, Ben-Shalom and River, 1984) has suggested the following process. Flavedo flakes are liquified at pH 4.5 with the aid of pectolytic enzymes and cellulase at 40°C for 6 hours. The terpene d-limonene is added to the mixture in order to absorb the readily extracted carotenoids. After first centrifuging the solid particles, the liquid is again centrifuged to remove the highly coloured d-limonene, which now contains 75% of the carotenoids present in the flavedo. The oil phase is dewaxed, concentrated and purified. A 3.6% carotenoid solution can be achieved by this method.

Extraction of colour from dried orange peel. An extraction process for colour from dried citrus peel invented by the late M. Koffler has been carried out on an industrial scale in Israel during several production seasons. Peel, rag and flakes are shredded and treated with lime to a pH of 8.0–9.0. They are then pressed and dried in a rotating drier with a directly heated drum. Temperatures are carefully controlled to avoid overheating the peel. The dried peels are stored in a cool silo and extracted as soon as possible to prevent loss of carotenoids. Extraction is carried out in a commercial extractor, as used in the oil industry, with light petroleum ether mixed with alcohol. The extract is concentrated under vacuum and d-limonene is added to the highly concentrated colour solution. The petroleum ether is removed by evaporation at under 50°C and a concentrated carotenoid solution in d-limonene of 4000 p.p.m. results.

8.3.2 Extraction of colour from grapes

Anthocyanins are present in many dark coloured fruits, such as red grapes, blue and red berries, dark cherries and some tropical fruits. Commercially, it is generally the wastes of the grape juice and wine industries that are used for extraction. Anthocyanin powder and concentrated anthocyanin solution (typically 4 g/100 g) are prepared for the food industry. Recently a very detailed work on the distribution and analyses of anthocyanins in fruit and vegetables has been published (Mazza and Miniati, 1993).

The anthocyanins present in food products are very sensitive to heat, metals, pH and air; therefore, it is not easy to produce attractive products without adding natural colour. The concentration of anthocyanins in grapes is dependent on the temperature during ripening. The optimal temperature is about 20°C, and a temperature of 35°C will prevent colour development. The most abundant anthocyanin in grapes is malvidin 3-acetylglucoside. Its concentration rises steadily during the ripening process, then declines slightly at the end of the process. Soil, water supply, light and fertilisers are additional factors affecting the development of anthocyanins in grapes.

The anthocyanins in grapes are extracted mainly from the skin of dark coloured varieties after the juice is pressed. About 20% of the total fruit weight is skins, seed and stems. In the extraction of the pomace, a diluted solution of sulphur dioxide is used, the same as in the maceration process of grapes in the wine industry. The sulphur dioxide acts as an antioxidant in the process of extraction. After colour extraction is completed, the solution is filtered and concentrated under vacuum causing a nearly complete evaporation of the sulphur dioxide.

8.4 Apple waste treatment

Apple juice production consists of several stages:

1. Grinding to produce the mash
2. Enzyme treatment
3. Pressing
4. Juice treatment
5. Pomace disposal or its use.

The pomace resulting from pressing mashed whole apples contains about 20–30% dry matter, 1.5–2.5% pectin and 10–20% carbohydrates.

As large quantities of apple juice are produced in the USA and Europe, the problem of waste disposal is a significant one. There are several ways of solving this:

- Feed the pomace directly to animals. Little investment is involved but accurate planning of disposal during a 24-hour period has to be undertaken. Because the pomace is poor in protein, it must be mixed with additional nutrients.
- Ensilage the pomace to give a relatively stable product. Typical analyses are given by Steingass and Haussner (1984).
- Dry the pomace for pectin production (see section 8.5).
- Dry the pomace for feed; this is only possible in large plants.

An additional use of apple waste as dietary fibre and cake filling has been suggested. During recent years, the liquifying of the apple mash with special enzymes and the secondary extraction of the pomace with water have reduced the amount of waste material from apple juice production.

8.5 Production of pectin

Pectin represents an important component of the waste of two commercially important fruit types, citrus and apple. In the waste of citrus juice production, the pectin content is up to 4% of the fresh weight and in apples it is up to 2%.

Having acquired the technology to extract purified pectin solutions enabling their use in many food products, pectin has become a highly valuable by-product of the fruit juice industry. Pectin, being a natural plant component, has its use not only in food products but also in nutritional and medical applications with the great advantage of being legally acceptable. In recent decades, with increasing emphasis on the favourable functions of pectin in human metabolism, its economic value has increased considerably.

Many aspects of pectin have been studied, particularly regarding its role in plants throughout their life cycle and in products prepared with them. Also many publications have appeared about its interaction with the pectolytic enzymes present naturally in plants and in the microorganisms that degrade them.

8.5.1 Characterisation of pectin and pectolytic enzymes in plants

The main pectin-containing tissue in the plant is the parenchyma. These cells are surrounded by primary walls and between them there is a middle lamella, both containing pectic substances. In the primary cell wall, the pectin, together with other materials, represents encrusting substances in between the cellulose fibrils, while the middle lamella (the main component) holds together the structure of the plant tissues (Schwimmer, 1981).

These pectic substances are polymers of d-galacturonic acid, partially esterified with methanol and coupled by $\alpha(1-4)$-glycosidic links (Kertesz, 1951; Doesburg, 1965). The degree of polymerisation is so high that these pectic substances are designated as protopectin and are insoluble in water. The elaborate structure of pectin is summarised by Rolin and De Vries (1990) and Pilnik and Voragen (1991). The synthesis of pectin during cell formation is discussed by Northcote (1986).

Doesburg (1965) summarising the data reported in the literature on the pectin content of apples during the harvest season observed a nearly constant level with a slight decrease at the end of the season. Kertesz

(1951) reported that in English apples the percentage of pectin based on fresh weight decreases during ripening and then becomes constant. Sinclair (1961) concludes that in orange peel during maturation and until senescence, the percentage of total anhydrogalacturonic acid on a dry weight basis first decreases and then levels off, but the relation of water-soluble pectin to total pectic substances rises until the peak of maturity and then gradually decreases.

Many data regarding the pectin content of citrus and apple can be found in the literature. Generally speaking, the pectin content in apple residue is about half of that in citrus peels. Within the genus citrus, lemon, lime and pomelo are the richest in pectin and mandarins the poorest; oranges and grapefruit have an intermediate position. As far as orange varieties are concerned, experience shows that Valencia is richer than Shamuti and Washington Navel, but no figures have been published.

8.5.2 *Pectin enzymes in plants used for production of pectin*

The pectolytic enzyme present in all plant materials used for pectin production is pectin methylesterase, which removes the methoxyl group from the carboxyl function. To a lesser extent polygalacturonase, abundant in fungi, is present in higher plants but not in apple, and it is doubtful if it occurs in any citrus fruit. Polygalacturonase hydrolyses glycosidic bonds next to free carboxyl groups. Two additional enzymes splitting glycosidic bonds by beta-elimination are pectate lyase, mainly present in bacteria, and pectin lyase. The distribution of pectolytic enzymes is summarised by Pilnik and Voragen (1991).

The pectin manufacturer who needs to reach the required range of esterification in the final product is obliged to control the pectin methylesterase activity in the raw material. Whilst natural polygalacturonase is not a problem, the presence of additional polygalacturonase and pectin lyase from fungi, yeasts or bacteria, infecting decaying fruit, especially towards the end of the season, or from microbial contamination during production, will cause concern. The activity of these enzymes of microbial origin may result in a loss of yield, deterioration of quality or, in extreme cases, even a complete failure of production.

8.5.3 *Commercial pectins and their production*

To render the pectin suitable for use in different applications, the molecule has to be made soluble. This demands some shortening of the chemical chain. The degree of esterification (DE) also will determine the characteristics of the final product and in the raw material this is very high and has to be adjusted in the process. The two processes, shortening of the polygalacturonic acid chain and partial de-esterification, might interfere with each other. Success is based on the art of the manufacturer who has to control the process to achieve the correct balance of these two requirements and obtain the maximum yield of the product. Obviously each manufacturer keeps the details of his process secret and, therefore, little has been published on this subject.

The commercial pectins consist of two basic groups according to their DE, with several variations in each group:

Pectins with DE above 50% (high methoxyl pectin, HMP). These pectins form gels only in the presence of a soluble solids content (TSS) above about 57% or by replacing part of the water with an organic solvent at a pH below 3.5. These pectins are prepared by direct extraction, which in the traditional process is under acid conditions at elevated temperature, and are sequentially precipitated.

Pectins with DE below 50% (low methoxyl pectin, LMP). These pectins form gels at lower TSS, depending upon the degree of methoxylation and require calcium ions for jellification. They are prepared either by direct extraction under more severe conditions or by modifying pectin previously extracted as HMP.

Each group of pectins includes various types, differing in chain length and degree of methoxylation and consequently have different characteristics. Each pectin manufacturer controls the production conditions in order to obtain the required type. Although they give the same name to each type, for instance, slow set, medium rapid set, these are similar but not exactly identical. In addition, pectin is made from a natural raw material that is influenced by climatic conditions, which vary each season and area, by agricultural practices, and by conditions at fruit harvest and transport to the factory. Besides scrupulous control of production, the manufacturer must constantly counterbalance the influence of these factors. Even so, some slight differences may occur, which will have to be compensated for by the final user.

One of the major problems of pectin manufacture is the disposal of effluent. Pectin manufacture increases the value of the waste material from fruit juice production but its traditional manufacturing process demands considerable quantities of water and its effluents often contain undesirable materials, such as soluble carbohydrates and neutralised acids.

Commercial preparation and sources of raw material for the production of pectin. Kertesz (1951) describes 50 different plants where details of the pectin content have been published in the literature. In order to obtain the desired quality of final product, the protopectin should be esterified with methanol and from the commercial point of view the percentage of protopectin should be high and sufficient quantities of raw material must be available at an economic price. In western Europe and America, raw material with these characteristics is limited to the residues of apples and citrus fruits after juice processing. During the 1980s, the use of apple pomace has decreased considerably, with citrus peels being richer in pectin and more readily available in sufficient quantities. Also, changes in the processing of apples by the use of liquifying enzymes has rendered the pomace unsuitable for pectin production and an increase in value of apple pomace for cattle feed has diminished its attractiveness for pectin production. However, some food manufacturers still prefer apple pectin for some uses where specific properties are required, probably because of the presence of certain starchy fractions remaining in the final pectin. Citrus peels treated with enzymes are not suitable for pectin production. Other raw materials not originating from fruit juice manufacture have at certain times been used for pectin production. Examples are the residues from sugar beet processing and sunflower seed heads. As the pectin from these sources is partly esterified with acetyl groups, and sunflower contains undesirable terpenes, these raw materials are not popular for this purpose. Recently, however, a plant in western Europe began to produce low-methoxyl pectin from sunflower seed heads. It is also believed that a specific variety of water melon is used in Russia as raw material for the production of pectin.

Because the period of waste availability after juice processing is limited and the plants of pectin manufacturers are often not located close to juice plants, a considerable proportion of the raw material has to be dried.

Certain points are essential in the preparation of dried peels for pectin production.

1. The limitation, if not inactivation, of pectin esterase activity in the raw material and the avoidance of pectolytic enzymes from any type of microorganism is essential, particularly if the raw material is transported long distances to the drying plant.
2. The waste should be shredded and washed in order to remove soluble carbohydrates and to make it compressible. A counter-current washing system may be used.
3. Drying at the lowest possible temperature and storage under dry conditions allowed forced ventilation with cooled air prevents overheating by chemical reactions, which might cause spontaneous combustion (Rebeck, 1990).

A pectin plant situated next to a fruit juice plant has the advantage of being able to avoid the need for drying during the season. Using this procedure for citrus peels it has been possible to increase the yield by 30–40% and improve the quality. The solution of pectin produced in this way is more cloudy.

Manufacture of pectin. Until recently, the usual process in western countries was to mix raw materials with acidified water for a certain time at elevated temperature. Each manufacturer selects his own parameters for preparing the different types of pectin. Although the principles are similar, the time of extraction may vary between 30 minutes and 45 hours, the temperature may vary between 40°C and 92°C and also the conditions of acidification differ. The change in the pectin molecule and its dissolution may be affected in one single or two separate steps, and in the latter procedure the ratio between the plant material and the acidified water may vary. The type and concentration of acid used will differ, not only according to the properties of the raw material and type of pectin required but also according to economic and environmental considerations. After separation of the crude pectin extract from the solid particles, it is clarified either by passing through filter presses with a filter aid in one or two steps, by centrifugation or by combining the two procedures.

In recent years Golubey and Gubanov of the Interbios company, in co-operation with the Russian Institute of Food Industry, have developed an essentially different process, using ultrasonics to split the protopectin. This process has been called 'cold pectin technology'. The process has successfully passed the pilot stage and the first industrial trial, including filtration through a membrane system and spray drying, is now planned. If this method proves technologically and economically successful, it will contribute much to the solution of the environmental problems and may be a serious competitor to the traditional way of production.

Two alternative processes for the production of high-methoxyl pectin have been used until recently in the western world. One is based on concentrating the purified pectin extract in a multiple-phase evaporator and precipating the pectin with alcohol. The other precipitates the pectin from the purified extract with an aluminium salt. The different steps necessary to secure the final product for these two processes are represented in a flow sheet in Figure 8.5. In this figure the various stages where waste is created are indicated as (A), (B) and (C) for solid, liquid and slurry consistency, respectively. These waste materials create serious environmental problems and, in spite of huge investments to try to solve them, have caused the closure of some pectin plants. The solid waste may be used as fodder with or without such treatment, as may be required locally.

The flow chart shown in Figure 8.5 has also been used for the production of low methoxyl pectin, but with increasing time, acidity and different conditions of precipitation. Usually it is more common to start with high methoxyl pectin precipitated either with alcohol or with aluminium salts and de-esterification by acid or base in alcohol at reduced temperature. By de-esterifying with ammonia, the methyl ester is partially substituted by amide groups. This type of low-methoxy pectin is called amidated. A maximum of 25% amidation is permitted by law in some countries.

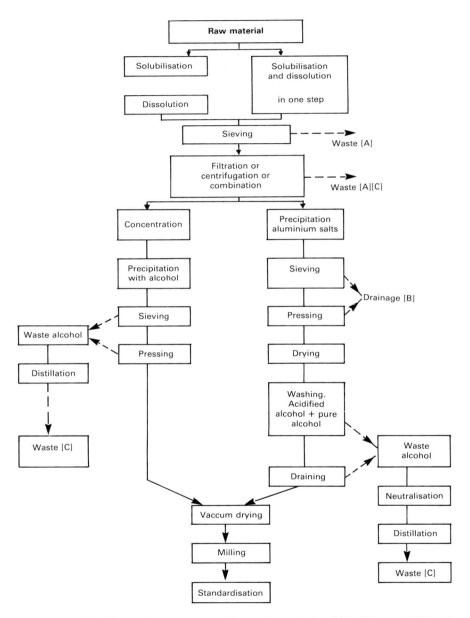

Figure 8.5 A flow diagram for two processes for pectin production. [A] solid waste; [B] liquid waste; [C] slurry.

The pectins produced will have different gelatinisation powers, according to their nature, the storage conditions of the raw material and the production process. A number of instruments can be used for measuring the 'jelly strength'. As these devices do not measure the same parameter, it is impossible to correlate their results. The various methods have been discussed by Crandall and Wicker (1986) and a short description is presented by Rolin and De Vries (1990). Up to now the universally used method is the USA–SAG method (Committee for Pectin Standardisation, 1959). The standard for international trade is 150 grade USA–SAG, which means that under standard conditions one part of pectin causes jellification of 150 parts of refined sucrose at standard elasticity.

8.5.4 Different types of pectin and their application

The generally accepted nomenclature of pectic substances is that formulated by Baker et al. (1944). A slight change has been proposed by Doesburg (1965).

The main application of pectin in the food industry is as a gelling agent. The commercial types of pectin are divided into two main groups depending on the mechanism of gelatinisation: those with esterification above 50%, which gelatinise only at TSS levels above 55% and in a restricted range of pH, and those with esterification below 50%, which gelatinise at a wider range of TSS and pH values in the presence of calcium ions. The theoretical explanation of the jellification process has been recently discussed by Crandall and Wicker (1986).

High-methoxyl pectins. Within this group, all other conditions being kept constant, the rapidity of setting is determined by the degree of esterification. The higher the percentage of esterification, within the range 57–82%, the more rapid the setting. Usually three types are offered in this range: 'Slow set', 'Medium rapid set' and 'Rapid set', but sometimes 'Ultra rapid set' is also available. The different manufacturers do not indicate in their catalogues exactly the same range of esterification for these types. The rapidity of setting depends also upon the concentration of TSS, and if this exceeds about 74% even Slow-set pectin might cause too rapid setting. Therefore, pectin manufacturers offer pectin mixed with inorganic and/or organic buffer salts and/or partially amidated pectin which retard the speed of setting.

High-methoxyl pectins are also used for increasing the viscosity of liquids. The higher the degree of esterification, the higher the viscosity obtained. This characteristic can be used for stabilising drinks and increasing mouthfeel, with the advantage that the additive is natural fruit component. Another application is their ability to emulsify essential and, especially, vegetable oils. It is successfully used to prevent curdling of acidified milk products and, therefore, facilitates pasteurisation of a mixture of milk and fruit juices.

Low-methoxyl pectins. At a lower degree of esterification, pectin can cause jellification at lower total solids content. Accordingly different types of low-methoxy pectin are offered for the production of dietary jams with different levels of sugar content, and milk products and aspics, especially those with low sugar content. Theoretically, jellification can be obtained without any sugar present, but in practice a 20% sugar content is the lowest limit because below this syneresis occurs.

Because low-methoxy pectins are produced by at least two different methods and several variations can be obtained by mixing with other organic polymers and other ingredients, the types are not so uniform. For any specific application, one of the manufacturers most specialised in this field should be consulted. After long discussions and much laboratory work, the safety of amidated low-methoxy pectins has been approved by the international health authorities.

Intermediate pectin. Pectin with a degree of methoxylation between 50 and 55% takes an intermediate position, combining to some extent the characteristics of both high- and low-methoxyl pectin. In the presence of calcium ions, it is less sensitive to the exact total solids content and pH than high-methoxy pectin and, therefore, has found its application in household pectin where a strict control of the cooking time to obtain jellification of jam is not vital.

Mixtures of pectin with other polymers. Pectin may be mixed with other gelling or thickening agents, but usually the nature of the admixed material is not disclosed. Morris (1990) and Toft, Grasdalen and Smidsrod (1986) have summarised the data concerning the characteristics of gel caused by mixing pectin and alginate compared with that of each component alone.

8.5.5 Application of pectin in medicine and nutrition

In spite of some difficulties in formal definition, pectin is now considered to be within the group of fibre substances (Baker, 1980) and all theories regarding this group apply to pectin. The beneficial effects of pectin in cases of diarrhoea have been known for many years. Several medicaments used against this ailment contain pectin. Also in the diets of patients, even babies, suffering from diarrhoea, plants rich in pectin are used.

Diverse aspects of the interaction between pectin and several metal ions have been studied. One of these is the detoxification of poisonous metals, such as lead and arsenic, which are bound by the pectin molecule (Kertesz, 1951). A marked decrease in heavy metals in the blood has been observed after the administration of pectin solutions and consequently pectin is regularly provided to people working under conditions liable to cause poisoning with harmful metals (Russian Institute of Food Industry, private communication). Behall and Reiser (1986) summarising the literature on absorption of metals in the present of pectin, report contradictory results. However, two patents were granted in the USA in 1967 where pectin was used as a carrier of metal ions for improved incorporation in the human body. One concerns bismuth (US patent 3,306,819) used in the treatment of gastrointestinal diseases and the other is iron (US patent 3,324,109) used to cure its deficiency (Lawrence, 1973). Cerda (1988), however, states that pectin has no influence on absorbtion of iron, and, undoubtedly, Behall and Reiser (1986) are correct in suggesting that further research is required.

Since the early 1970s much attention has been paid to the level of cholesterol in human blood, and consequently the possible influence of pectin has become the subject of much research. Positive results may be of commercial value for the producer.

Cerda (1988) claims that the daily intake of 15 g pectin causes the cholesterol level in human blood to decrease, and Baig and Cerda (1980) explain it by the formation of insoluble complexes of pectin with low-density lipoprotein. Unfortunately, this has been proved only *in vitro*. In most cases in which more than 6 g pectin per day was administered, the cholesterol level decreased (Behall and Reiser, (1986).

Attention has been paid also to the hypoglycaemic effect of pectin and Behall and Reiser (1986) summarise the papers published between 1976 and 1984. In about 65 of the cases, administering pectin decreased the glucose level in blood.

References

Anon (1987). *Requirements of Dairy Cattle*, 5th edn. National Academy of Sciences, Washington, DC.
Baig, M.M. and Cerda, J.J. (1980). *Citrus Nutrition and Quality* (ed. Nagy, S. and Attaway, J.A.) ACS Symposium Series 143, p. 25. American Chemical Society, Washington, DC.
Baker, R.A. (1980). *Citrus Nutrition and Quality* (ed. Nagy, S. and Attaway, J.A.) ACS Symposium Series 143, p. 109. American Chemical Society, Washington, DC.
Baker, R.A. (1993). *IFT Annual Meeting – Book of Abstracts*, pp. 43, 153, Chicago.
Behall, K. and Reiser, S. (1986). *Chemistry and Function of Pectin* (ed. Fishman, M.L. and Jen, J.J.) ACS Symposium Series 310, p. 248. American Chemical Society, Washington, DC.
Ben-Shalom, N. and Pinto, R (1986). *Lebensmittel Wissenschaft und Technologie*, **19**, 158.
Cerda, J.J. (1988). *Confructa*, **32**(1), 6.
Cohen, E., Sharon, R., Volman, L. *et al.* (1984). *Journal of Food Science*, **49**(4), 987.
Crandall, P.G. and Wicker, L. (1986). *Chemistry and Function of Pectin* (ed. Fishman, M.L. and Jen, J.J.) ACS Symposium Series 310, p. 88. American Chemical Society, Washington, DC.
Doesburg, J.J. (1965). *I.B.V.T. Communication No 25*, Wageningen.

Ferguson, R.R. and Fox, K.I. (1978). *Trans. Citrus Eng. Conf.*, Vol. 24, p. 23.
Gross J. (1977). *Citrus Science and Technology*, Vol. 2 (ed. Nagy, S., Shaw, P.E. and Veldhuis, M.K.), p. 302. Avi Publishing, Westport, CT.
Hasegawa, S, Brewster, L.C. and Maier, V.P. (1973). *Journal of Food Science*, **38**(7), 1153.
Kanner, J., Ben-Shalom, N., and Shomer, I. (1982). *Lebensmittel Wissenschaft und Technologie*, **15**, 348.
Kanner, J., Ben-Shalom, N., and River, O. (1984). *Proceedings of the International Federation of Fruit Juice Producers*, **18**, 219.
Kertesz, Z.I. (1951). *The Pectic Substances.* Interscience, New York.
Kesterson, J.W. and Braddock, R.J. (1973). *Food Technology*, **27**(2), 50.
Lawrence, A.A. (1973). *Edible Gums and Related Substances.* S. Noyes Data Corporation, Park Ridge, New York.
Mazza, G. and Miniati, E. (1993). *Anthocyanins in Fruits, Vegetables and Grains.* CRC Press, Boca Raton, FL.
Morris, A.R. (1990). *Food Gels* (ed. Harris, P.), p. 344. Elsevier Applied Science, London.
Northcote, D.H. (1986). *Chemistry and Function of Pectins* (ed. Fishman, M.L. and Jen, J.J.), ACS Symposium Series 310, p. 134. American Chemical Society, Washington, DC.
Pilnik, W. and Voragen, G.J. (1991). *Food Enzymology*, Vol. 1 (ed. Fox, P.F.), p. 303. Elsevier Applied Science, London.
Ragonese, C. (1991). *Fluessiges Obst*, **58**(5), 222.
Rebeck, H.M. (1995). *Production and Packaging of Non-carbonated Fruit Juices and Fruit Beverages* (ed. Ashurst, P.R.), Blackie A&P, Glasgow.
Redd, J.B. and Hendrix, C.M. Jr. (1993). *Fruit Juice Processing Technology* (eds Nagy, S., Chin, Shu Shen and Shaw, P.E.). AgScience Inc, Auburndale, FL.
Robbins, R.C. (1980). *Citrus Nutrition and Quality* (ed. Nagy, S. and Attaway, J.A.), ACS Symposium Series 143, p. 43. American Chemical Society, Washington, DC.
Rolin, C. and de Vries, J. (1990). *Food Gels* (ed. Harris, P.), p. 401. Elsevier Applied Science, London.
Schwimmer, S. (1981). *Source Book of Food Enzymology*, Ch. 29, p. 511. Avi Sourcebook and Handbook Series, Avi Publishing, Westport, CT.
Sinclair, W.B. (1961). *The Orange, its Biochemistry and Physiology*, University of California.
Steingass, H. and Haussner, A. (1988). *Confructa Studien*, 96.
Toft, K., Grasdalen, H. and Smidsrod, O. (1986). *Chemistry and Function of Pectin* (ed. Fishman, M.L. and Jen, J.J.), p. 117. ACS Symposium Series 310, American Chemical Society, Washington, DC.
Wilson, C.W. Bisset, O.W. and Berry, R.E. (1971). *Journal of Food Science*, **36**(6), 1033.
Wright, J. (1995). *Food Flavourings* (ed. Ashurst P.R.), p. 24. Blackie A&P, Glasgow.

Further reading

Baker, G.L., Joseph, G.H., Kertesz, Z.I. et al. (1944). *Chemical Engineering News*, **22**, 105.
Bruemmer, R.B. (1976). *Proceedings Florida State Horticultural Society*, **89**, 191.
Cohn, R. (1985). *Confructa Studien*, **29**(3), 178.
Griffiths, F.P. and Lime, B.J. (1959). *Food Technology*, **13**(7), 430.
Hofsommer, H.J., Fischer-Ayloff-Cook, K.P. and Radcke, H.J. (1991). *Fluessiges Obst*, **58**(2), 62.
Horowitz, R.M. (1961). *The Orange, its Biochemistry and Physiology* (ed. Sinclair, W.B.). University of California.
IFT Committee on Pectin Standardisation (1959). Final report. *Food Technology*, **13**(8), 496.
Kimball, D. (1991). *Citrus Processing, Quality Control and Technology*. Avi Book, Van Nostrand Reinhold, New York.
Shaw, P.E. and Wilson, C.W. (1983). *Journal of Food Science*, **48**(2), 646.
Ting, S.V. and Hendrickson, R. (1969). *Food Technology*, **23**(7), 947.

9 Water supplies, effluent disposal and other environmental considerations
M.J.V. WAYMAN

9.1 Introduction

As water is the principal component of fruit and fruit products, it is surprising that only relatively recently its sourcing and disposal have been given due consideration. The more settled climate of past decades has given way to more erratic weather patterns world-wide. This change has far-reaching implications for crop husbandry and, not least, for environmental concerns. In periods of drought or areas of low rainfall, the recycling, or at least the further use of water, even if only for irrigation, has become pressing.

Attention to the usage and disposal of water is, therefore, claiming more management time and investment in technology than ever before. This is not without its benefits. Improved control can be exercised over both the quality and the economics of production. Indifferent attention has, in the past, resulted in off-flavours, spoilage and frequent extravagance in the volumes of water used (particularly in primary processes). Improved technologies have become available. These enable easy close monitoring of key parameters and, by means of appropriate software, the use of this additional data for the better management of the water sector.

The advent of environmental assessment philosophies such as British Standard (BS) 7750 implies that interest in the ways that companies use raw materials, natural resources and energy will increase. This interest will come not only from environmental auditors, but also from purchasers of produced goods, shareholders and, not least, potential investors.

There is an international dimension, too. Long-distance distribution has resulted in the development of more generally agreed standards, arguable though some of these still are. The European Union is a clear example, with its series of published International Standards Organisation (ISO) objectives. Compliance with these, rather than with local or national standards, will increasingly become the norm.

9.2 Water sourcing

Fruit processing has historically made use of considerable amounts of water. Rising costs of acquisition and disposal have prompted an interest in reducing this use. Hydraulic discharge flumes have been replaced by conveyor belts, and solid wastes collected dry instead of being flushed to drain. The trend from canning to freezing has not had as much impact in this area as might have been expected.

Town water requires the minimum of capital investment on the part of the fruit processor. The supply is usually dependable and its characteristics not subject to great variation. As the costs are not under the control of the consumer they can rise at rates that threaten profitability.

In order to minimise spoilage from handling and delays occurring in transportation, many fruit processing plants are located in more remote areas where the contributing crops are grown. Here, recourse has to be made to a local river, borehole or spring. Provided that appropriate necessary treatment is applied, all these can be made entirely satisfactory for use.

River water varies the most widely in composition as a rule, and so calls for the longest sequence of unit treatments to render it suitable for use.

Unless the water source is a major one, storage will be required at some point. This may be before any treatment, combined with a settlement/clarifying stage or later on as partially or fully treated water. The capacity of such a reservoir can readily be calculated. Due allowances may be made for extreme demands (firefighting, large-scale washdown), and transfer pumps rated accordingly.

9.3 Primary treatment

Only rarely can water be used as it is acquired. Some boreholes and conduited springs may seem to qualify, but even if innocent of unwanted components, the water from them may still need adjustments in regard to pH value, alkalinity and biological stabilisation.

The range of characteristics requiring attention before raw waters comply with fruit industry standards includes the removal of suspended solids, compounds conferring colour, taste and/or smell, dissolved minerals, metal ions, salts and, not least, microorganisms: animal, vegetable, fungal, bacteria and viral.

The minimum standard to be aimed for is that of potable water. In this connection, a most useful and readable introduction will be found in Volume 1 of *Guidelines for Drinking-water Quality*, published by the World Health Organization.

9.3.1 Screening

Water intakes from rivers should be carefully sited so as to minimise the ingress of silt and floating matter. Suitable screens will keep out larger debris and aquatic life. Fine strainers fitted to pump suction pipes will keep out smaller particles. Vigilance has to be exercised to maintain these in good order and self-cleaning devices are available for these duties. Downstream grit channels can easily be arranged to allow sand and small stones to drop out under gravity.

9.3.2 Colour removal

Surface waters are often brownish as the result of vegetative breakdown products, such as humic acid derivatives and/or iron species. The standard response to this is to apply a coagulant treatment followed, for example, by a polyelectrolyte flocculant. The extensive internal surface area thus produced adsorbs coloured elements, clay, chalk and so on. After settlement, a clear and bright water results. Iron and aluminium compounds are available in several forms: bench trials are employed to establish the best one and at what optimum dose. The selected treatment should be monitored so that adjustment can be made to cope with seasonal or other variations in raw-water quality.

Borehole water may require particular treatment to deal with dissolved metals (iron, manganese) and/or anionic species (bicarbonate, sulphide). Depending upon the amounts present, chemical dosing, aeration towers and catalytic filters are used to deal with these components (Figure 9.1). All reagents and materials in contact with the water should be approved 'for use with potable water'.

9.3.3 Adjustment of pH

Fruit products will tolerate only a limited range of pH values before changes of hue or colour become evident, or before hydrolysis results in irreversible impairment of flavour, vitamin content or physical structure. Natural waters can be acidic or alkaline: towns water is usually well corrected to around neutrality. Local primary treatment, with metal salts and/or lime, may result in a water with a pH value that needs further

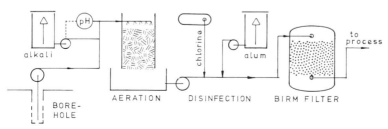

Figure 9.1 A bore-hole water treatment plant in West Malaysia. Disinfection uses chlorine.

adjustment before use. Granted a retention time of a few minutes and fully mixed conditions, a pH meter-controller can add reagent acid or alkali automatically to achieve and maintain any desired target range.

pH adjustment is the usual way in which dissolved toxic metals may be removed by precipitation as their hydrated oxides. Some metals (e.g. nickel) require high pH values to achieve minimum solubility; others again may be present in anionic forms (e.g. chromate, plumbate), which are not precipitatable.

9.3.4 Filtration

Unless extended time is available, settlement will not usually result in exhaustive removal of suspended solids. Induced currents and thermal eddies will always conspire to give residual solids of a few parts-per-million. Filtration can reduce these to whatever low limits may be specified for particular applications. Municipal waterworks use gravity sand-beds but a more appropriate industrial option is that of pressure-filters of various configurations. These can include conventional fixed beds of gravel, sand and sometimes anthracite, depth-wound cartridges and supported membranes. All filters eventually become charged with accumulated solids, which are disposed of either by back-washing or by changing the filtration element. It is, therefore, desirable to install filters in duplex working mode. On occasions, filters can become breeding places for microorganisms and, therefore, the means of sterilising or brine washing should be included.

9.3.5 Carbon adsorption

Certain organic compounds, present in only trace amounts, impart unacceptable odours and flavours to water. These stem largely from human sources: materials of construction (for example, bitumen joint sealants), or direct pollution of an aquifer or water course. The latter may be contaminated by the upstream outfall of a mine or effluent treatment works.

Techniques are now available to detect and monitor sub-microgram concentrations of solvents, oils, phenols, hormones and pesticides. If not dealt with at this point in the treatment sequence, later chlorination may exacerbate problems (e.g. convert phenols to chlorinated derivatives.

Most of the above can be removed satisfactorily by passing the water through a fixed bed of activated carbon. Granular forms are derived from coal or a range of vegetable materials: coconut shells and particular softwoods (vine wood). All these give charcoals having open microstructures and are sometimes treated chemically to impart mechanical or enhanced adsorptive properties.

As with sand filtration, means have to be engineered for occasional back-washing, bed expansion and/or regeneration. It may be possible to effect the last *in situ*, or the carbon may have to be removed for stoving. Again, duplex units are indicated.

9.3.6 Primary disinfection

Water may appear to be of good quality but still carries bacteria, viruses, yeasts and mould spores. These would carry through to the fruit products and lodge in pipework to become a source of chronic infection.

Conventional biocidal agents include halogen species (chlorine, bromine, iodine), silver, oxidants (peroxide, ozone, ultraviolet irradiation) and reducing agents (sulphur dioxide). On grounds of applicability and cost, chlorine is the preferred initial option. Its most effective germicidal derivative is hypochlorous acid (HOCl). The proportion of this present in any situation is related to the net pH value. Above pH 9, it has dissociated and is ineffective; below pH 7.5, acceptable activity is secured.

Chlorine may be applied as the gaseous element dissolved in recirculated water, as sodium or calcium hypochlorite, or as part of an organic carrier, such as trichloroisocyanuric acid (or its sodium salt). The choice will depend upon the scale and engineering of a particular operation. Standard plant is available for any demand.

Sufficient reagent is added to maintain a small 'chlorine residual' at the farthest point of distribution. For normal drinking, cooking and washing purposes in temperate countries, this may be only 0.2 mg/litre. In the tropics, a higher figure (0.5 mg/litre) is desirable (Figure 9.2). Water treated in this way will be acceptable for all less critical processes: fruit preparation and incorporation into fruit beverages and juices. Further treatment may be required for certain particular applications. Residual chlorine can, however, also cause tainting in some fruits and fruit products.

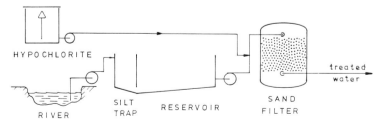

Figure 9.2 A river-water treatment plant in northern Nigeria using a reservoir, sand filtration and hypochlorite disinfection.

9.4 Secondary treatment

9.4.1 Boiler feedwater

Considerable quantities of low-pressure steam and high-pressure hot water (HPHW) are used in fruit processing. If boilers are to continue to function efficiently, then precautions must be taken to protect them from corrosion and scaling. Often, it is sufficient to add appropriate chemicals to the feedwater; phosphates, tannins, phosphonates, alkalis, etc. Sometimes, partial softening or demineralisation will have to be employed. There should be regular monitoring of the boiler contents, and prompt and sufficient blowing-down practised when indicated to control the concentration of dissolved salts. Condensate recovery is usually undertaken: protection of this system from corrosion is often by adding a filming amine (e.g. octadecylamine) to the feed-water.

9.4.2 Cooling water

The requirements here parallel those for boiler feed water. Precautions have to be taken to prevent scaling and corrosion. Additionally, biocides must be used to control fouling and consequent blockage of pipework and heat exchangers.

The health hazard posed by Legionnaires' disease has prompted moves towards air-cooled systems and, sometimes, to the installation of refrigerated chilled-water facilities. As a rule, the suppliers of proprietary treatment chemicals will advise and monitor the use of appropriate reagents.

9.4.3 Water for bottle-washing

Hard water leaves unacceptable films of dried salts on glassware and at least the final rinsing should be performed with soft or softened water. The base-exchange process is simple and relatively cheap to operate with calcium and magnesium ions in the water being replaced by sodium. Subsequent heating does not then throw any precipitates. Regeneration of the ion-exchange resin is achieved by a means of saturated solution of salt (brine), with calcium and magnesium chlorides being discarded to waste. This treatment also allows the reduced use of surfactants (detergents).

9.4.4 Water for fruit dressing

Virtually all fruit as received will be contaminated to some extent with soil, plant debris, juice, pesticide and/or other husbandry residues. After primary sorting, the greater part of this foreign matter may be removed in appropriate washing equipment. Medium- to high-pressure water is sprayed through arrays of nozzles so as to reach all surfaces. Potable quality water is acceptable for this operation.

If the cost of water is significant, or if it is in short supply, then some degree of recovery may be practised, via a screen and pressure filter combinations, perhaps. In any case, it would be desirable to meter the water used, so that proper audits may be carried out periodically.

For cut fruit, additional precautions must be taken to guard against spoilage. Water for this purpose is chlorinated to a higher level; a residual of between 25 and 30 mg/l is often adopted but the product must be monitored to ensure freedom from taints.

9.4.5 Water for process use

The characteristics required of process water may be determined by the particular use to which it is to be put. Canning, fruit-juice production, jams, wine-making or pharmaceutical preparations exhibit very different internal physico-chemical environments. A water satisfactory for one product may well not suit another. pH value, alkalinity and salts content can all bear noticeably on flavour and colour. Tolerances to these may vary not only from product to product, but also between fruit, fruit varieties and between fruit grown on different types of soil or under different conditions. The subtleties of wines demonstrate this very well.

Trials should be conducted to determine if a particular water supply is acceptable for the process use intended. If some unwanted effect becomes apparent, then it will be necessary to investigate and pursue an appropriate remedy. Depending upon the problem, the action required may be a simple chemical adjustment – of pH value, for example – or some more radical further treatment may be necessary.

In some areas, the mains water supply may be over-chlorinated to an extent that affects the fruit/products adversely. The colour may be impaired and/or the flavour diminished. In these circumstances, the water must be presented to a de-chlorinating filter, based on activated carbon. Water treated in this way loses its residual protection against infection. If the dechlorination is practised at the point of use, this will be acceptable. Otherwise, the risk may be under-written by the later inclusion of a local disinfecting stage. Almost always, this will be by ultraviolet irradiation. A high-actinic discharge tube is mounted within a silica sheathing tube, placed in turn in an in-flow casing (Figure 9.3). Means can be included for monitoring the continuing performance of the unit.

9.4.6 Water for special applications

For certain uses, the process water may be required to comply with various national or international pharmaceutical standards. Generally, these relate

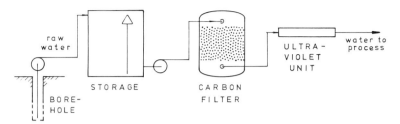

Figure 9.3 Water for process use: soft drinks manufacture in South Wales. Water is stored, passed through activated charcoal and irradiated with ultraviolet light.

to the dissolved salts content and the microbial count. The former may be controlled by a specific property (conductivity), a net parameter (hardness, dissolved solids), or a specific limitation (chloride, nitrate). The contributing minerals may be removed down to the specified levels by several means: by a sufficient degree of deionisation (demineralisation), by a membrane process (reverse osmosis) or by distillation either alone or in some combination. The choice of system will depend upon the size of the demand for this water and on the particular standard to be achieved (Table 9.1). Ion-exchange needs regenerant chemicals and effluent disposal. Reverse osmosis discards considerable amounts of waste water and is not exhaustive. Distillation uses the most energy. A typical treatment train for BP water comprises break tank, reverse osmosis, mixed-bed ion-exchange, ultraviolet disinfection unit and sub-micron cartridge filter.

9.4.7 Cleaning-in-place (CIP)

Whatever system is installed for the treatment of supply water prior to process use, there may come to be occasions when some, at least, of the installed equipment will require cleaning. Deposits of suspended matter, hardness scale or even biofilm may accumulate progressively or suddenly and unnoticed. Although plant should not be dismantled wantonly, periodic inspection of key parts is advisable.

CIP procedures will depend, of course, on the nature of whatever accumulations are discovered. Formulations are available from specialist supply companies and may use acids, alkalis, surfactants and biocides, alone or in combination.

The design stage of a water treatment installation should allow, therefore, for appropriate inspection, entry and drain-down points for CIP purposes.

Table 9.1 Typical water quality standards

Category	Standard	pH value	TDS	S/S	T.H.	M/O	Conductivity
Potable	WHO	6.5 to 8.5	500	1.0	500	100	5
BP	BS 3978	6.0 to 8.5	1.0	0.1	0.1	10	0.1

9.5 Effluent planning

Whilst the collection and disposal of surface (rain) water is usually given proper consideration when laying out a factory site, too often comparable arrangements for process effluents are neglected. At worst, it may be assumed that a single drainage scheme will suffice for both. This is inviting trouble. Extremes of rainfall can overload any finite effluent plant and result in the escape of untreated wastes to the receiving water course. High oxygen-demanding material can then asphyxiate aquatic life. It may also affect downstream uses for irrigation or potable supplies to others.

Factory lay-outs should be arranged so that receipt of fruit and primary handling are as remote as possible from finished goods production and storage. This may be achieved in a notional way by the provision of bunds or other means whereby separation can be rigorously observed.

The natural slope of the terrain is also important. Floor gullies, drains, catch-pits and sumps all need adequate relative inverts for proper operation. Falls need to be steeper if solids are likely to be present. Due allowance should always be made for maintenance access. This may include rodding and/or flushing with jetted water.

Underground pipework should be buried at such depths as to comply with local by-laws and be adequately marked, both on site and on drainage lay-out drawings. Information regarding nominal bore, invert and materials of construction should be included on the latter.

Underneath roads, yards and other trafficked areas, buried pipes will need additional protection, usually by being sheathed in concrete.

Ideally all these factors should be considered at initial site planning. The later addition and construction of effluent collection systems can be very much more difficult and costly.

9.5.1 Segregation

Although a single system for effluent and rain-water drainage may appear easier conceptually (and certainly cheaper), in practice it suffers from inflexibility and operational problems. Segregation can facilitate recovery of energy, of materials and of water itself.

For instance, sensible heat from condensate and used bottle-washing rinses may easily be imparted to incoming cold water by appropriate heat exchangers. Water discarded by reverse osmosis may be used for other, less critical processes.

9.5.2 Effluent transfer

It often happens that the ideal of a fully gravitational flow system cannot readily be attained. There will then have to be sundry collection sumps with associated pumps to transfer the waste-waters to the distant treatment facilities. Every sump should be provided with at least a simple screen, and the pumps (duty and stand-by) fitted with strainers. Debris of all sorts finds its way into factory drains: some can cause considerable trouble if they lodge in critical places.

Preferably local collection points should have local devices to separate fruit solids from the waste stream. These take the form of cylindrical or parabolic screens of proprietary design. Mechanical clearance is often included. Macerators are generally not to be approved as they serve to increase the load of dissolved oxygen-demanding material which burdens later treatment processes. The transfer pumps can be controlled automatically, by conductance or by other level-monitoring system. The last can also have a flood-level alarm and be interlinked with the main control console.

Pipe runs should be chosen intelligently to avoid too many changes of direction. Swept bends are to be preferred to elbows. Care should be taken to size the pipework appropriately. If too large, the lower velocity may allow silt and heavier solids to settle and cause obstruction. In temperate countries, there may be a requirement for heat tracing and lagging against freezing. Adequate support, physical protection and labelling are essential in all cases.

Pipework and fittings may most conveniently be of polymeric material. Unplasticised polyvinylchloride (uPVC) is the cheapest overall; acrylonitrile–butadiene–styrene co-polymer (ABS) is stronger and has better thermal resistance; polypropylene resists solvent effects best but requires welding installation. For heated water, of course, stainless steel will usually be specified.

Throughout the site, the greatest care should be exercised to ensure that all fruit juice wastes are comprehensively collected and routed only by the designated drainage collection and transfer system. Leakages and spillages to the foundational sub-soil will result in contaminated ground-water, with potentially grave consequences:

- acids may corrode concrete and reinforcing steelwork
- the fruit juice may ferment and loosen the soil structure
- solutes may migrate into local aquifers, bore holes or water courses
- the organic material may form a substrate for sulphate-reducing bacteria which draw their oxygen supply from sulphate (present in concrete). The

consequences of such activity can be the evolution of hydrogen sulphide gas. This is toxic at the parts-per-million level and has the insidious property of deadening the human olfactory sense. If the smell is overlooked, collapse and possibly death may ensue.

Lengthy and costly remediation may involve replacements of civil foundation structures, as not only will the withdrawal of sulphate weaken concrete, but the hydrogen sulphide itself rapidly corrodes steel and cast iron.

9.5.3 Effluent reception

Some of the effluents will be fairly consistent in flow and composition. Others, for example, machinery wash-downs, will be intermittent and may carry strong chemical loads (sugar, caustic soda). Measures to handle these in the reception and/or transfer stages are essential to the economic operation of any subsequent treatment system. The reception tanks for all streams should be made large enough to provide not only sufficient flow and load balancing, but also spare capacity to allow short-term problems with treatment facilities to be addressed.

On an existing site, a survey will soon indicate requirements. This should be based on reputable samples drawn from representative points and over sufficient time to embrace the range of variations likely to be experienced. For a projected fruit factory, best estimates should be made and an agreed notional effluent schedule compiled. This should also allow for any foreseeable future developments.

9.5.4 Treatment objectives

These may be less or more onerous depending upon the site and on the subsequent destination of the used and treated water. At the very least, some limitations will be imposed on the pH value, the solids content and on the demand for oxygen exhibited by the net effluent. Increasingly, maximum levels are also being set to govern other parameters, such as colour and temperature.

As with the supply water, standard reference methods are published, which should be used by the controlling analytical laboratory. Occasional short-cuts may be adopted for internal use (e.g. Brix measurement for sugar content).

Discharge consent results may be categorised or banded according to the nature of the receiving water course or other disposal. A small works adjacent to a major river or the sea may, by virtue of the dilution available, have relatively easy limits to attain (see Table 9.2).

Table 9.2 Typical discharge limits

Parameter	Water course	Local drainage
pH value	6 to 9	6 to 10
BOD5	Max. 20 mg/l	–
COD	–	Max. 600 mg/l
Suspended solids	Max. 30 mg/l	Max. 400 mg/l

9.6 Effluent characterisation

For the purposes of effluent treatment, the nature of the processing waste is not as important as its effects.

The pH value can have a critical influence on chemical and biochemical reactions. In addition to the effluent pH values, the corresponding acidities or alkalinities must be determined because of buffering effects. Fruit acids in particular can be responsible for giving very misleading pH results.

The measurement of pH values is best made using a pH meter with a special electrode, as colour indicators do not work well with fruit wastes.

9.6.1 Suspended solids

The assessment of this is not straightforward. At the simplest level, it is the weight of insoluble material per volume of effluent. A sample is filtered through glass-fibre sheet and dried to constant weight or for a specified time at an agreed temperature, usually 105°C, before weighing. In practice, it may be important to distinguish between 'settleable' solids, related to a certain time scale, and non-settling, colloidal matter. Also, besides the weight/volume ratio, there is occasionally a requirement for the volume/volume information for the same solids.

9.6.2 Oxygen demand

Apart from a small minority of specialised microorganisms, all life on earth requires access to oxygen in some form or another in order to function. In the aquatic environment, this is available in solution at a concentration in equilibrium with the atmosphere. In cold waters, this is about 10 mg/litre. In warmer conditions, this falls to 7 mg/litre or so.

In a water course, anything that competes for this oxygen threatens whatever life is already established there. Examples include inorganic species having reducing properties, such as ferrous iron, thiosulphate and bisulphite, and soluble organic compounds. The measure of the threat which all these represent is termed their oxygen demand.

The original analytical method for determining this property uses a culture of microorganisms and takes five days of incubation; it is known as measuring the biological oxygen demand or BOD5.

Two, more rapid, techniques use chemicals as reagents. Potassium permanganate for the oxygen absorbed (OA) or permanganate value (PV) test, and chromic acid for the chemical oxygen demand (COD) test. Each of these methods has its uses. There are no universal interconversion factors, though for particular effluents, some rough and ready relationships may be established by experience (Table 9.2).

Because there are some compounds that react partially or not at all with these oxidising agents, a further analysis is sometimes performed for total organic carbon (TOC). This involves pyrolysing the sample in a stream of oxygen and measuring the carbon dioxide evolved. This test has the merit of avoiding errors caused by the inclusion of inorganic reducing substances as well as functioning in the presence of inhibitors to the foregoing methods.

9.6.3 Other parameters

As suggested above, there may be particular local interest in other characteristics such as temperature and colour. The quantative measurement of the latter is not easy, even using instrumentation, and various contingent standards are in use.

9.6.4 Effluent monitoring

A data base of effluent records is invaluable. An appropriate programme of sampling, testing and flow recording, allied to production figures, will enable optimisation of the use of water and the minimisation of waste. pH values, temperatures and flow rates are easily logged automatically and downloaded when required to computer spreadsheets. Abnormalities and trends can then easily be recognised and responded to.

9.7 Effluent treatment

Fruit processing effluents can present a wide range of characteristics. A treatment process or sequence of processes appropriate to one may not suit others. The several discharge consent conditions should be considered item by item against each scheduled contributor and an initial table of mismatches drawn up.

It will then be necessary to produce integrated design variants to minimise the engineering ultimately to be installed: opportunities for mutual treatment may well present themselves, alkali peeling liquor to neutralise citrus juices, for example, or chlorinated water to satisfy part of an oxygen demand.

It may be wise to hold back relatively clean streams until the end of another treatment sequence, so as to take advantage of the dilution thus available. The possibility of partial or side-stream treatment should also be considered. Some processes result in a product water so well within some limiting parameter that only a corresponding portion needs to be treated. An example here might be pressure filtration to trim suspended solids down to a finite residual figure.

Eventually, an appropriate overall scheme will be arrived at. As a rule, this will comprise a sequence of unit processes, each dedicated to the limited adjustment or removal of particular parameters. The selection and specification of these processes are a matter of judgement based on the effluent schedule, the discharge limitations and on any incidental recoveries which it may prove feasible to practise.

9.7.1 Solids removal

In a gravity-fed system, all manner of plant debris may be expected. The simplest way of removing this is by passing the flow through a run-down, parabolic screen. Larger installations will require mechanically brushed or raked devices of varying sophistication. Collected material should be removed regularly to approved disposal.

Finer material is generally reluctant to settle. It may be helpful, particularly if colloids are present, to insert here a chemical addition stage. The judicious use of ferric and/or aluminium salts, perhaps in conjunction with a polyelectrolyte flocculant, can coagulate and absorb a significant proportion of the polluting load. Laboratory treatability studies will show if this step would be worthwhile or not. Both ferric chloride and ferric sulphate are available: the former tends to be slightly more expensive but can give superior results. It should be noted that all these are acidic in nature and that some corresponding pH correction may be needed.

Alternatively, induced or assisted flotation could be intalled at this point. Microbubbles are generated within a tank containing the effluent, by electrical or mechanical means. In the latter case, the technique is known as 'dissolved air flotation' (DAF). The bubbles nucleate on dispersed particles and increase their buoyancy so that migration to the surface is accelerated. A mousse-like layer builds up and is removed progressively by a mechanical skimmer. The air is derived from a small compressor and contacted with a percentage of recirculated water at a pressure of some 6 bar. The amount of air capable of being dissolved at this pressure is about ten times ambient.

9.7.2 pH adjustment

As this is directly related to the presence of acids and alkalis, this will depend on the type and variety of fruit being processed, the end-product in view and the use of any ancillary chemical (e.g. for cleaning, sterilising, demineralisation and so on).

The pH of each effluent will have an initial value which may change with time, as the result of chemical reaction (hydrolysis or oxidation) or biological activity (fermentation, enzymic transformation). It may be necessary to adjust the pH value not only initially but also subsequently, to allow a process to proceed optimally. Bio-oxidation performs best in slightly alkaline conditions, between pH 7.0 and 8.0. Acidic fruit juices must, therefore, be neutralised before being presented to this treatment. Further, the degradation itself results in the production of acid (carboxylic) residues. There will, therefore, be a corresponding need for continuing pH control.

The measuring electrode of the pH meter is usually made of glass, but if fluorides are present (from the surface etching of aluminium fruit juice/beverage cans), then an antimony electrode has to be substituted. It is important to locate the pH probes so that they measure representatively. They require regular inspection and cleaning, together with occasional recalibration, if they are to continue to function properly. Since pH reference cells have a finite life, it is advisable to replace these elements annually as a matter of course.

The neutralising reagents most readily used are sulphuric acid and sodium hydroxide. The latter may conveniently be acquired as a concentrated solution (caustic soda liquor). Both can be purchased in IBC tanks containing 1000 litres, from which they can be dispensed as needed. Both these reagents should be diluted to between 5 and 10% strength; this can be carried out automatically in a day-tank of appropriate size.

The neutralisation reaction is virtually instantaneous, but because of statistical 'leakage', it is usual to allow a retention time for this stage of at least 15 minutes. The transfer of reagent can be by actuated valve, if inverts permit, otherwise pumps may be used. For larger installations, a ring main with branches to solenoid valves is favoured. Duty, boost and alarm functions should always be included. The tank or pit should be

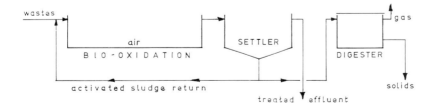

Figure 9.4 Conventional bio-oxidation treatment system.

provided with an electric stirrer matched to the duty. Direct-coupled, marine propeller types are suitable for vessels up to 5 cubic metres; above this, a geared turbine will be required.

9.7.3 Biological oxygen demand (BOD)

This is by far the most significant of effluent properties needing attention. It will require the greatest capital investment and entail the highest running costs. Fruit processing wastes are made up of considerable amounts of biodegradable solids and dissolved matter: fruit acids and proteins, carbohydrates, sugars, colloids and esters. All these contribute to the net BOD loading.

Naturally occurring microorganisms are used to absorb and digest the dissolved materials from the effluent; these are converted into solid and mineral forms (biomass), which can then be separated by conventional means. Most of the carbon is lost to the atmosphere as carbon dioxide, while organically bound sulphur and nitrogen are converted to sulphate and either nitrate or nitrogen itself.

Optimum conditions for bio-oxidation can be achieved in several ways, depending upon the scale of the exercise, the type of effluent being handled, the availability of space and, not least, the target quality required for the fully treated discharge. The metabolic requirements are oxygen, appropriate temperature and pH ranges and the necessary retention times for contact with the effluent, for subsequent stabilisation and for removal by settlement.

It is essential, too, that an acceptable balance be maintained between the nutrients necessary for growth (ammoniacal nitrogen, phosphates and trace minerals) and the carbon content of the organic material present. In round figures, the balance should be BOD:N:P in the ratio of 100:5:1. Fruit-processing effluents are usually deficient in both nitrogen and phosphates to the extent that supplementation will be required. Appropriate amounts of stock solution are added by dosing pump, the rate being monitored by the site control laboratory and altered from time to time as found necessary.

All biocidal chemical compounds *must* be excluded and this should be understood by everyone on site. These include herbicides applied to pathways etc., insecticides used to control drosophilae, and janitorial disinfectants. There are also machinery maintenance proprietaries (solvents, sterilisers, lubricant/hydraulic silicones), timber preservatives, rodent control preparations and fuel. Alternative means of disposal should be provided for all these: washing them 'down the drain' may so interfere with biological treatment that the plant could take weeks to recover fully.

9.8 Forms of biological treatment

9.8.1 *The activated sludge process*

This is the simplest form of bio-oxidation. The effluent stream is brought into contact with an appropriate culture of microorganisms and the whole is agitated with air. Organic matter is digested and converted; after several hours, this process reaches completion and the effluent is said to have been stabilised. The solids component augmented thereby is then allowed to separate under the influence of gravity, and the treated effluent discharged as permitted.

There are three principal ways in which air can be introduced:

- Mechanical agitation of the surface: various designs of rotor are available, which are mounted in some way just below the surface. Electric motors drive each at high speed. Water is picked up and dispersed over the adjacent area at the same time as air is driven into the body of the effluent. Some concomitant noise is generated. It is not usually possible to modulate mechanical aerators; that is, they are either on or off.
- Tank aeration: a remote blower supplies a header-and-lateral manifold mounted across the bottom of the tank or pit. At spaced intervals are fitted diffusers, cap-pieces, bubble guns, helix tubes or other devices for breaking up the air and inducing a degree of turbulence. By including valves in the distribution system, it is possible to economise when less than the full amount of air is needed.
- Venturis: this simple system is useful for smaller installations. A proportion of the effluent is recycled through a suitably specified venturi unit. Atmospheric air is drawn in and intimately mixed with the flow which then rejoins the aeration tank. Energy requirements are competitive with the previous methods, and the effluent enjoys lateral mixing.

Although batch treatment is sometimes practised, most activated sludge plants are designed on a continuous, flow-through basis. Dimensions are based on between 6 and 8 hours' passage. A proportion of the separated biomass, the activated sludge, is recirculated to seed the influent stream. Since aeration costs money, however, some form of control is usual. Measurements of dissolved oxygen (DO) are made at strategic points and the system adjusted accordingly.

To avoid the capital costs of a large plant that could lie idle for much of the year, experiments have been conducted with smaller plants which employ pure oxygen and are used when required to meet short-term, seasonal demands. The oxygen can be 'switched on', as it were, only when needed to provide a contrived extra capability. Very careful attention needs to be given to the cost effectiveness of each candidate plant.

There are three established developments of the activated sludge process, each offering certain refinements and advantages.

In the 'extended aeration' method, the retention time is lengthened to a day or more. This gives a plant that occupies more space, of course, but which copes better with variations of load and gives improved performance via exhaustive digestion and solids consolidation (Figure 9.4).

If the long channel above is made into a closed loop, an 'oxidation ditch' is formed. This is often oval, rather than circular, and it is provided with some way of inducing the effluent to flow. Horizontally mounted brushes or paddles are used. Air is introduced by these and/or a surface agitator or blower. A section of the ditch is provided with a baffle, behind which surplus solids/biomass settle out and is removed at intervals. It works well with consistent balanced effluents, but once installed its range of adjustment is very limited. If the retention time is long enough, reliable results can be achieved.

The third development 'contact stabilisation' is more engineered but more controllable and adaptable. The process is separated into compartments dedicated to the different stages. Incoming effluent is mixed immediately with a considerable proportion of well-aerated sludge. This rapidly takes up the dissolved organic material. After a short time, the effluent passes to a settlement stage, from which the clarified portion runs directly to outfall. The settled sludge has not by then completed its digestion of the absorbed material, so is pumped to a separate tank for aeration. The latter can therefore be much smaller than for aeration of the entire flow, and this saves energy (Figure 9.5).

This version is efficient in its use of space and can cope well with peak loads. Operational variables include the rate of return of the sludge and its degree of aeration. Long-term sludge reduction can be included. A large installation will require full-time oversight, for sampling, testing and making corresponding adjustments to particular plant functions.

Attempts to improve the efficiency of the activated sludge process have been made from time to time by adding some dispersed substrate to the aeration stage. Sponges and magnetised particles are examples, both being

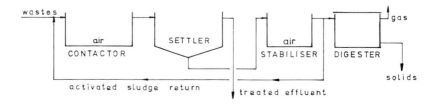

Figure 9.5 The contact stabilisation system.

readily removable by appropriate means. These have not been as effective as envisaged.

9.8.2 Percolating 'filters'

The activated sludge techniques depend upon the managed admixture of effluent, air and suspended biomass. The critical and limiting factor is the surface area of the biomass across which exchanges of organic material and oxygen take place. An alternative way of engineering this is to support the biomass on some fixed medium. Nineteenth century studies resulted in the development of beds of mineral material (clinker, slag or road-stone), on to the top surface of which effluent is trickled or sprayed. The beds may be circular in plan or rectangular. Biomass forms on the surface of the mineral, absorbs organic matter, grows and sloughs off from time to time for conventional separation. The rate of presentation must be related to the BOD loading and to the capacity of the installation. Trickling filters enjoy widespread use. They are frequently installed in tandem, with periodic alternation of the order in which they are used. Known as 'alternating double filtration' (ADF), this practice is capable of achieving very low residual levels of BOD. Although cheap to run, such filters do not adjust well to variations in load or balance, and are somewhat at the mercy of weather conditions.

9.8.3 High-rate filtration

The advent of specially designed plastics elements has allowed a more intensive use of space for biofiltration. A simple enclosed tower structure is filled with either random or ordered plastics modules, and the effluent distributed over the top layer. Biomass establishes itself on the exposed surfaces and grows as described previously. Because of the open nature of the filling media, the effluent travels rapidly down to a collection sump underneath. From here, some of it is recirculated, while the remainder runs forward to a settlement stage. Running and maintenance costs are very modest, and units have been installed that have been found to be of great value in a wide variety of situations.

This technology has been found in practice to cope well with high loadings and fluctuations. Its performance is relatively indifferent, however, being between 65 and 75% BOD removal per stage. It is usual, therefore, to place two towers in series to achieve some 90% removal. This may suffice for some situations, but more often some secondary treatment is called for. This can be either activated sludge (the contact stabilisation mode), or AD filtration.

9.8.4 Mechanical contacting

A more intensively engineered form of high-rate filter is the 'biocontactor'. This comprises an array of circular elements, perhaps 3 or 4 metres in diameter, mounted along a slowly rotated shaft. The elements may be either rigid corrugated discs or caging enclosing random-fill plastic packing pieces. Their lower portions are immersed in a trough containing the effluent. Biomass develops on the elements and acts upon dissolved organic matter as previously described. Biocontactors offer the best utilisation of space, if that is at a premium.

9.9 Tertiary treatment

9.9.1 Filtration

If very low discharge limits have to be attained, further 'polishing' may be necessary. Gravity sand filters can be used to filter out residual solids, as in water-works practice. For organic traces, if sufficient land is available, constructed wetlands or 'reedbeds' may commend themselves. There are several variations of these, the most popular being the horizontal flow, root-zone format. Shallow trenches are lined with an impermeable membrane and filled with large aggregate. Inlet and outlet arrangements are made using gabions. Seedlings of the common reed, *Phragmites australis*, are planted at 1 metre intervals. This plant has the unique property of releasing oxygen from its extensive root/rhizome system. Biomass builds up on the aggregate and functions in combination with the reeds. Removal of up to 98% of the applied BOD has been reported. Two growing seasons are needed for full efficiency to be reached, but, thereafter, many years usage may be expected, with very little required in the way of maintenance.

9.9.2 Solids removal and disposal

The settlement stages associated with the foregoing depend upon the effluent being allowed quiescent conditions so that the upwards flow vector of the effluent flow is less than the downwards fall rate of solids particles present. This is achieved through the dimensional ratios and other aspects of their design. Vertical, horizontal and radial patterns are all well known, the last two being applicable more to larger duties. Fittings to settlers include inlet distribution launders and still-wells, outlet weirs, intermediate baffles and scrapers of half or full-bridge design.

A more rapid and complete clarification is often achieved if a small dose of polyelectrolyte is added just prior to settlement. This practice can help any later de-watering exercise and is inexpensive to institute and to run.

Accumulated solids can be digested and greatly reduced in volume by anaerobic digestion, but it is doubtful if the investment in the necessary equipment could be justified unless it was in year-round use, and/or if the resultant products (for example, methane) had sufficient market value.

Most crudely, surplus activated slurry can simply be sprayed on to whatever land is available. Since the consequent odour may be detectable a mile or more away, this practice is rapidly becoming unacceptable. Drying beds are often used: moisture is lost by a combination of seepage through the soil and by evaporation.

It may well be preferred, however, particularly on larger sites, to de-water the slurry mechanically. For this, the filter-press is a robust and reliable device. A horizontal stack of plates and frames is interleaved with an appropriate filter cloth, usually woven polypropylene twill. The slurry is pumped in at high pressure to result eventually in a 'cake' 15 to 25 mm thick having a solids content of between 15 and 30%, and of spadable consistency. Ultimate disposal may be to land distribution or, less usually, to incineration.

9.9.3 Recovery and re-use options

Attempts have been made from time to time to utilise waste fruit tissue and surplus biomass as a culture medium for growing single-cell protein, for incorporation in animal feedstuff. Low world protein prices have tended to render such attempts commercially unattractive. There is considerable potential, however, in drier regions, for the further use of spent water for irrigation purposes. Proper consideration must be given, of course, to crops appropriate to the situation and season(s).

9.10 Environmental auditing

This term implies not only to the straightforward quantification of the uses of natural resources and utilities, but also to judgements as to whether these uses have been optimised to achieve minimum environmental impact.

9.10.1 Baseline assessment

For a new fruit processing plant, comparisons of its later performance in regard to the environment at large will be made with reference to the parameters achieved prior to construction. It is not always easy to anticipate which of these may come to be of public interest.

Enquiries should be made of parties having an actual or potential interest in the fruit processing enterprise. These will include the supplier or suppliers of process and mains water (which may come from different sources) and the authorities regulating effluent and atmospheric discharges, and solid wastes disposal. Advice as to prevailing legislation should be sought also from the Health and Safety Executive, Food Industry Inspectorate, local planning authority and any relevant trade body. A further sector to be informed, if not consulted, is that embracing community, amenity and political interests.

Limited surveys covering disposals, fumes, noise, traffic and so on should be drawn up with the knowledge, at least, of the above. Ideally, to maintain objectivity, such work should be undertaken by an outside specialist company.

9.10.2 Period review

At suitable intervals, determined by experience and discussion, the baseline surveys should be re-visited and key elements repeated. Significant changes will show straightaway: more gradual trends will become apparent only as data builds up over longer periods of time. Graphical representations are always helpful and statistical analysis sometimes reveals information not otherwise readily discerned.

9.10.3 Quality and environmental standards

After a long period of development, during which individual companies experimented with their own schemes, the British Standards Institution published consolidated guidelines in its BS 5750 Standard (ISO 9000: EN 29000). Unfortunately, accreditation to this standard, or to any part of it, does not imply any objective compliances. Quality targets are entirely self-set. Of much greater import are the later BS 7750 Standard and the European Union's Eco-Management & Audit Scheme (EMAS). Assessments for awarding certification for these are obliged to take into account the effects of the business operation on the global environment, and to state whether these effects are having the least feasible impact. Considerations include the use of energy and water, disposal of service and manufacturing wastes (solid, liquid and gaseous) and the recycle potential of products and their packaging.

Environmental standards have largely been covered in the fore-going. It remains to reiterate that these are certain to become ever more limiting and more closely monitored. For new establishments, the regulating authorities will need to be assured of best environmental practice before overall planning permission is sought. In the long run, however, such are sure to return the initial investment made in them, as well as enhancing the environmentally responsible reputation of the operator.

Further reading

Water supplies and treatment

Camp, T.R. (1963). *Water and its Impurities*. Reinhold.
HMSO (1976 to 1992). *Index of Methods for the Examination of Waters etc*. HMSO, London.
Houghton, H.W. & McDonald, D. (1978). In *Developments in Soft Drinks Technology* (ed. Green, L.F.). Applied Science.
Sykes, G. (1965). *Disinfection: Theory and Practice*, 2nd edn. E. & F.N. Spon.

Waste treatment and disposal

American Chemical Society Division of Agriculture and Food Chemistry (1988). *Quality Factors of Fruits and Vegetables*. American Chemical Society, Washington, DC.
Besselievre, E.B. (1952). *Industrial Waste Treatment*. McGraw-Hill.
Gurnham, C.F. (1965). *Industrial Wastewater Control*. Academic Press.
Hayes, G.D. (1991). *Qualities in the Food Industry*. Department of Food Manufacturing, Manchester Polytechnic, UK.
Hevzka, A. & Booth, R.G. (ed.) (1981). *Food Industry Wastes: Disposal and Recovery*. Applied Science.
Hulme, A.C. (ed.) (1970). *The Biochemistry of Fruits and Their Products*, Vol. 1 (Vol II, 1971). Academic Press.
Koziorowski, B. & Kucharski, J. (1972). *Industrial Waste Disposal*, Pergamon.

Index

Acid
  ascorbic 29, 33, 36
  citric 32, 168
  malic 32, 168
adulteration of juice 94
air circulation rate 52
air separators 56
albedo 196
alginates 170
amino acids 23
angelica 185
anthocyanins 209–210
antioxidants 35
apple
  canned 145
  juice manufacture 77–79
  pie 147
  sauce 147
aroma 199, 205
aseptic packaging 75
  of juice 87
  of fruit 162
astringency 100
atmosphere generation 56
apricots, canned 148

bakery jams 181
berry juices 80
bilberries, canned 148
biological oxygen demand (BOD) 236
biological treatment of effluent 237
blackberries, canned 148
blackcurrants, canned 148
blanching 37, 138
blast freezing 159
boiling of jam 179
Botrytis cinerea 62
bottling
  of fruit 157
  of juice 88
breaking of tomatoes
  hot 193
  cold 193
brix 139
bruising 66
buffer salts 172
by-products 196

calcium 24
cannery hygiene 136

canning
  cookers 144
  juice 87
  process 142
cans 138
  closing 140
carbohydrates 23–24
carbon absorption for water 224
carbon dioxide 54, 59, 93
carbonated beverages 93
carbonyls 11
carotenoids 32, 36, 208–209
catalytic oxygen burner 56
cellulose 23
centrifugation 84, 91
chaptalisation 105
chemical preservatives 89
cherries
  canned 149
  cultivars 149
chestnut purée 171
chlorophyll 32
cider 98
  apple types 99
  chemistry 118
  cidre nouveau 111
  fermentation 103, 107, 113
  ice cider 111
  low alcohol 110
  maturation 119
  vintage 111
  white and black cider 110
citrus
  fruits 5
  oils 203
  peel 169, 185, 199, 206
  pulp 197
  rag 199
clarification
  of citrus oils 204
  of fruit wines 125
  of juice 84
climacteric 7, 41
colours 208
  removal from water 223
comminuted juices 206
concentration 89, 91, 200
  of citrus juice 75
controlled atmosphere 55
  generation 56

cooling/chilling 45
  coolrooms 53
  injury 45
  speed 58
cooling water 226
counter current extraction 78, 82, 197
crytosporidiosis 115
crystallisation 185
cultivars 2, 15, 71, 111
  apples for canning 145
  grapes for drying 186
  jam 167
  rhubarb for canning 156
currants 186

damsons, canned 155
dates 190
dealcoholisation of cider 110
debittering 200
decolourisation of cider 110
degreening 43
dehydrated citrus cells 198
dehydration/drying 36, 186–191
dietary fibre 24, 34, 211
diffusion extraction 78
diseases 61
  control of 62
disinfection of water 225
disorders 64

effluents 229–241
environmental auditing 241
  standards 242
enzyme 32, 85
  browning 79
  pectinase 74
esters 12
essence recovery 89
essential oils 199, 205
ethylene 8, 41, 42
evaporation of jam 179
evaporators 51
exhausting of cans 141

factory
  planning 229
  reception 137
fats 24–26
  in jam 169
fermentation 4
  double 117
  fruit wines 123
  secondary 114
  yeasts 113
field heat 47
figs 190
filling of cans 138

filtration
  of effluent 240
  of water 224
filtration sterilisation 89
fining of cider 109
flavedo 196, 209
flavonoids 33
flavour 191
  biosynthesis 10
  volatiles 75, 89
flowers, crystallised 185
fluidised bed freezing 159
fortified wines 127
freeze concentration 92
freezing 38
  of juice 80, 89
  of fruit 2, 158, 167
fruit
  cocktail, canned 151
  content in jam 171
  dried 190
  handling 40, 57
  salad, canned 151
  spirits, distillation 128
  wines 121
fungicides 63
  resistance 63

gas
  barriers 55
  analysis 57
gelling agents 169
ginger 184
glacé fruits 183
gooseberries, canned 149
grapefruit, canned 150
guava pulp 84

HACCP 176, 179
halogens 225
harvesting apples for cider 101
heat exchangers 86
hesperidin 17, 202
humidifiers 52
humidity 51
hydrocarbon fuel burner 56
hydrocooling 50

ingredients for jam 166
insect injury 66
irradiation 38

jams 165
jelly
  jams 171
  marmalade 175
juice drinks 93

INDEX

lactic acid bacteria 14
lactones 12
limonin 72, 200
liqueurs 129
liquid
  nitrogen 57
  freezing 160
loganberries, canned 152
low temperature injury 65
lycopene 29

mango pulp 81
manufacture of jams 176–182
marmalade 175
maturity 40, 71
  cider 119
  fruit wines 124
  standards 44
mechanical agitation of effluent 237
mechanical injury 66
methanol 128
milling of apples 76
minerals 24, 34
modified atmosphere 59
molasses 206

naringin 17, 32, 150
  production 202
near infra red spectroscopy 95
nectars 93
nitrogen 57
non-alcoholic fermented beverages 132
non-fermented juices 70
nutrient deficiency 65
nutrition 20
  composition 21–22

oils
  cold pressed 203
  concentrated 205
  distilled 205
oranges
  canned 152
  juice production 72–76
oxygen 54, 59
  demand 232

packaging 67
papaya purée 81
passionfruit juice 83
pasteurisation 75, 199
pastry for fruit pie 147
patulin 116
peaches, canned 154
pear
  canned 154
  juice 79
  perry 120

pectin 23, 74, 166, 169–170, 199
  medical application 218
  production 211–218
pectolytic enzymes 199
peeling 137
  of grapefruit 150
  of oranges 152
  of peaches 153
Penicillium sp. 62
perry 120
pH
  of effluent 234
  of water 223
phenols 12
pineapple
  canned 154
  juice 81
plums, canned 155
pomace 102, 210
pome fruits 3
potassium 24
preserves 165
pressing
  apples 77–78
  cider 102
proteins 23
prunes, canned 155
pulpwash 76, 197

quality 60, 70
  cider 109–110
  standards 3, 242

raisins 186
raspberries, canned 156
raw materials 135
recipes for jam 170–6
refrigeration 47
respiration 45, 56
  anaerobic 54
retorts for canning 144
reverse osmosis 92
rhubarb, canned 156
ripening 40, 42
  rooms 43
room freezing 159

Saccharomyces cerevisae 104, 113
Saccharomyces rouxii 75
safety hazards 43
screening of water 222
scrubbers 57
seaming of cans 140
selectivity ratio 59
shelf life 58, 75
snibbing 150
soft fruits 6
solar injury 65

solid pack
 apples   146
 rhubarb   157
solids removal   234
sorbitol   23, 35, 107
sparkling drinks
 ciders   117
 fruit wines   127
spin cooking   87
spiral belt freezing   159
spoilage
 of cider   114
 by microfungi   116
squash   206
stone fruit   5, 79
stone gum in plums   155
storage   40, 42, 137
 atmospheres   54
 cool   51
 drying   191
 frozen fruit   160
 jacketed room   52
strawberries, canned   157
strigging of blackcurrants   148
sucrose   168
sugar   168
sulphur dioxide   37, 103, 190
 in cider   111
 in jam   167
sultanas   186
suspended solids   232
syruping   139

taints   126
tank aeration of effluent   237
tartrate crystallisation   80

temperature
 control in cider   109, 113
 effects   45, 59
 frozen foods   161
terpenes   199, 206
thermal preservation of juice   86
tomato purée   191
treatment
 of effluent   231–241
 of water   222
turbidity   126

ultrafiltration   86, 92, 109

vacuum
 boiling   180
 cans   142
 cooling   50
 measurement   142
varieties   *see* cultivars
venturis   237
vitamins   29, 30, 31, 33

waste from apples   210
water   20
 boilers   226
 cleaning   228
 figs   20
 grapes   20
 processing   227
 strawberries   20
 supply   221
 tomatoes   20

yeasts   114, 116